Intermediate B Mathematics

Nancy McGraw

Bright Ideas Press, LLC
Cleveland, Ohio

Summer Solutions Intermediate B

Mathematics

Printed in the United States of America

ISBN 13: 978-1-934210-37-6
ISBN 10: 1-934210-37-4

Cover Design: Dan Mazzola
Editor: Kimberly A. Dambrogio

Instructions for Parents/Guardians

- *Summer Solutions* is an extension of the *Simple Solutions* approach being used by thousands of children in schools across the United States.
- The 30 Lessons included in each workbook are meant to review and reinforce the skills learned in the grade level just completed.
- The program is designed to be used 3 days per week for 10 weeks to ensure retention.
- Completing the book all at one time defeats the purpose of sustained practice over the summer break.
- Each book contains answers for each lesson.
- Each book also contains a "Who Knows?" drill and *Help Pages* which list vocabulary, solved examples, formulas, and measurement conversions.
- Lessons should be checked immediately for optimal feedback. Items that were difficult for students or done incorrectly should be resolved to ensure mastery.
- Adjust the use of the book to fit vacations. More lessons may have to be completed during the weeks before or following a family vacation.

Summer Solutions Intermediate B

Reviewed Skills

- Use of Basic Operations
- Place Value & Rounding
- Fraction Number Theory
- Addition & Subtraction of Fractions and Mixed Numbers
- Multiplication & Division of Fractions and Mixed Numbers
- Writing, Ordering & Rounding Decimals
- Addition & Subtraction of Decimals
- Multiplication & Division of Decimals
- Ratio and Proportion
- Percent
- Perimeter, Volume & Surface Area
- Geometry Formulas for Area (Triangle, Parallelogram, Circle, Trapezoid)
- Math Vocabulary
- Simple Measurement and Conversions
- Problem Solving

Help Pages begin on page 63.

Answers to Lessons begin on page 85.

Lesson #1

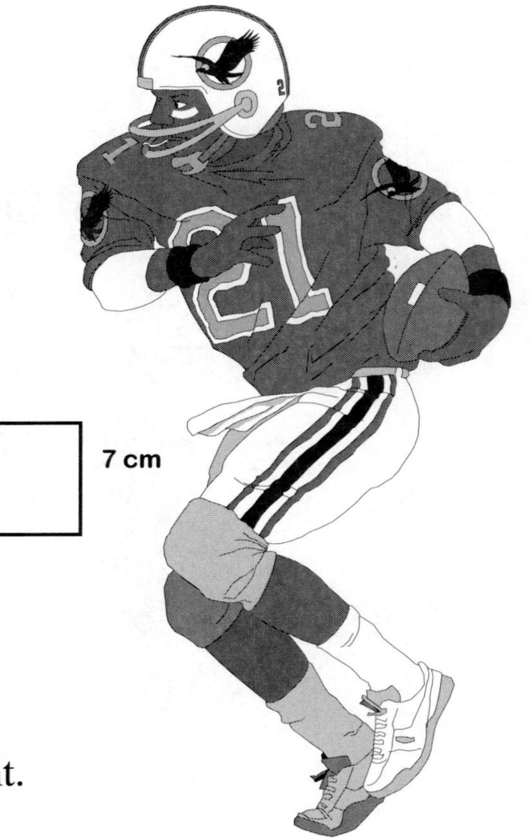

1. $6\frac{1}{9} - 3\frac{4}{9} = ?$
2. $0.034 \times 0.6 = ?$
3. Find the LCM of 15 and 25.
4. Write *0.7534* using words.
5. $50{,}000 - 25{,}781 = ?$
6. $8\frac{1}{5} + 6\frac{2}{3} = ?$
7. Find the area of this figure.
8. $65.45 \div 0.05 = ?$ 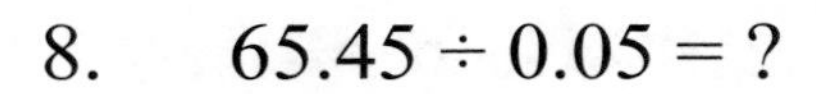
9. Find $\frac{2}{5}$ of 25. 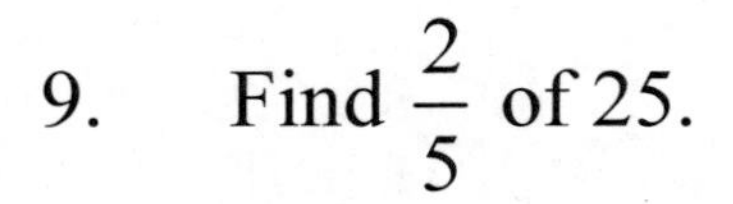
10. Write $\frac{7}{20}$ as a decimal and as a percent. 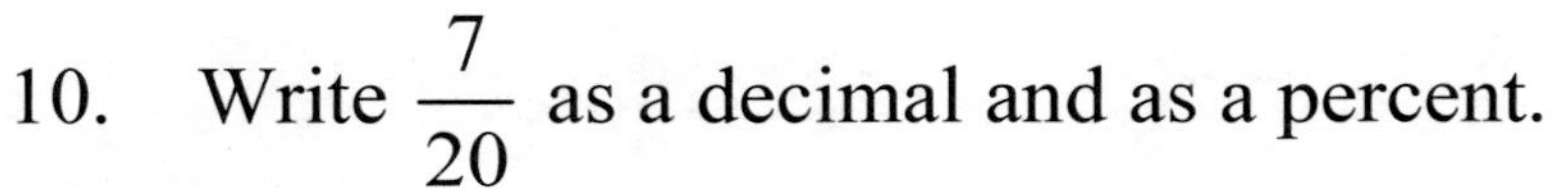
11. List the factors of *24*. 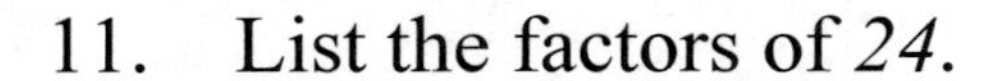
12. Solve for x. $\frac{3}{7} = \frac{x}{91}$
13. $87{,}345 + 98{,}673 = ?$
14. Round *47,864,210* to the nearest hundred thousand.
15. Put these decimals in order from least to greatest.

    0.3 0.035 0.53 0.503
16. Write $\frac{10}{15}$ in simplest form.
17. Find the mean of 65, 30 and 25.
18. What percent of 80 is 24?
19. Write the next number in the sequence. 46, 51, 56
20. A triangle with no sides congruent is a(n) ____________ triangle.

1.	2.	3.	4.
5.	6.	7.	8.
9.	10.	11.	12.
13.	14.	15.	16.
17.	18.	19.	20.

Lesson #2

1. $0.7 - 0.246 = ?$

2. Find $\frac{3}{7}$ of 21.

3. $0.848 \div 0.4 = ?$

4. Robb is in school 6 hours each day. What fraction of the day is Robb in school? What fraction of the day is he not in school?

5. Write the ratio *2:3* in another way.

6. $23{,}121 - 7{,}855 = ?$

7. If it is 5:30 now, what time will it be in 4 hours and 15 minutes?

8. Solve for x. $\frac{3}{4} = \frac{x}{64}$

9. Round *38,431,962* to the nearest million.

10. $7{,}864 \times 9 = ?$

11. A triangle with 2 sides congruent is a(n) ___________ triangle.

12. Find 70% of 60.

13. Jeff is 5 feet 5 inches tall. How many inches tall is he?

14. $346 + 987 = ?$

15. $2\frac{1}{2} \times 3\frac{1}{10} = ?$

16. Find the area of the triangle.

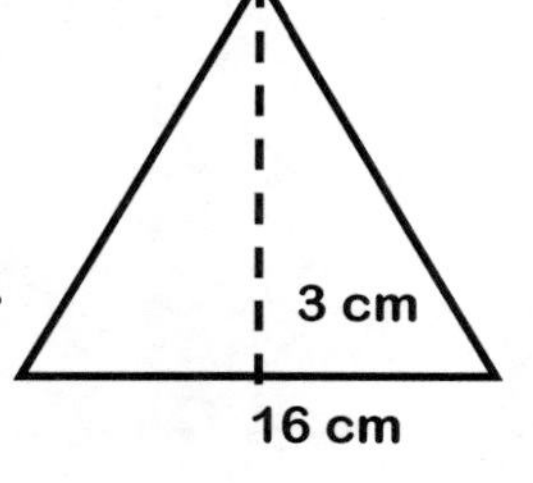

17. $\frac{5}{9} \bigcirc \frac{4}{7}$

18. $37 \times 85 = ?$

19. $\frac{6}{7} \times \frac{14}{20} = ?$

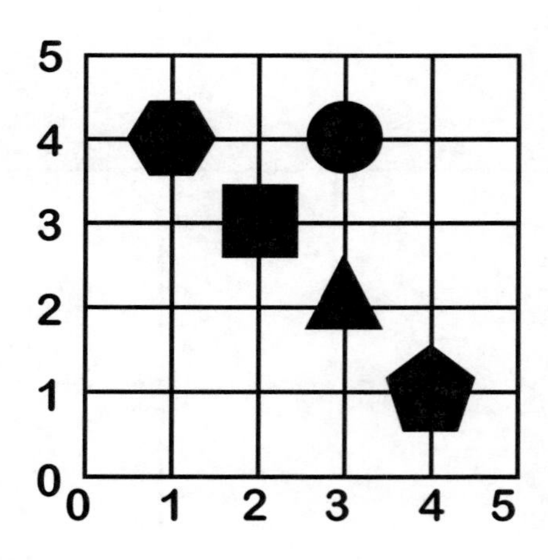

20. Name the shape located at (3, 2) on the grid. Give the pair of numbers that locates the hexagon.

1.	2.	3.	4.
5.	6.	7.	8.
9.	10.	11.	12.
13.	14.	15.	16.
17.	18.	19.	20.

Lesson #3

1. Write $\frac{3}{5}$ as a decimal and as a percent.
2. $432 \div 30 = ?$
3. $\frac{8}{10} \div \frac{2}{5} = ?$
4. $3{,}466 + 9{,}512 = ?$
5. Find the circumference of the circle.

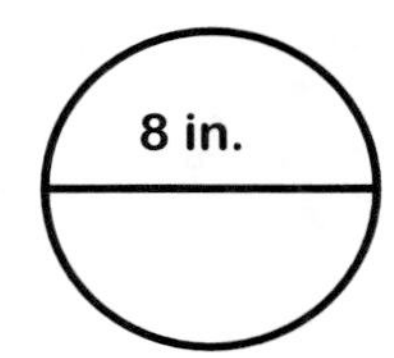

6. $4\frac{3}{4} + 8\frac{2}{5} = ?$
7. The area of a square is 64 ft^2. What is the length of each side?
8. How many grams are in a kilogram?
9. $656 \times 34 = ?$
10. If 5 yards of lace can make a dress, how many yards of lace are needed to make 25 dresses?
11. $2.7 \times 1.5 = ?$

12. Solve the proportion. $\frac{5}{6} = \frac{x}{90}$
13. Draw perpendicular lines.
14. $2{,}905 - 899 = ?$
15. $8 - 3\frac{3}{5} = ?$
16. How many feet are in 4 miles?
17. Write *0.5732* using words.
18. Find the GCF and the LCM of 16 and 20.
19. How many pounds are $3\frac{1}{2}$ tons?
20. How long, in inches, is the base of the pentagon?

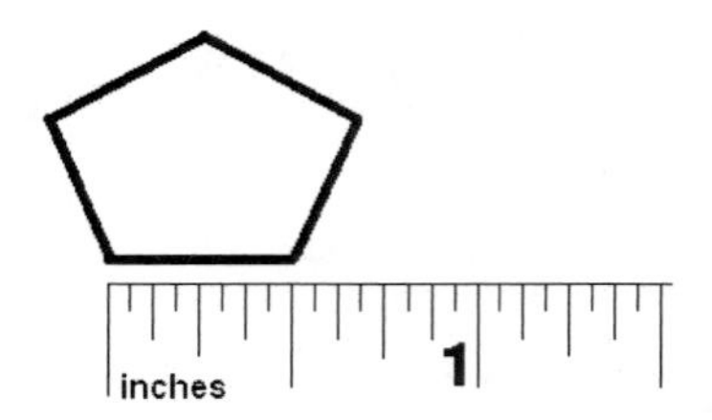

1.	2.	3.	4.
5.	6.	7.	8.
9.	10.	11.	12.
13.	14.	15.	16.
17.	18.	19.	20.

Lesson #4

1. How many years are in 7 decades?

2. $35.6 + 54.78 = ?$

3. $80{,}000 - 45{,}777 = ?$

4. Put these decimals in order from greatest to least.

   4.762 4.7 4.267 4.076

5. Find the median of 36, 14, 21, 56 and 10.

6. $957 \div 78 = ?$

7. $2\frac{1}{2} \times \frac{2}{5} = ?$

8. Find the area of the parallelogram.

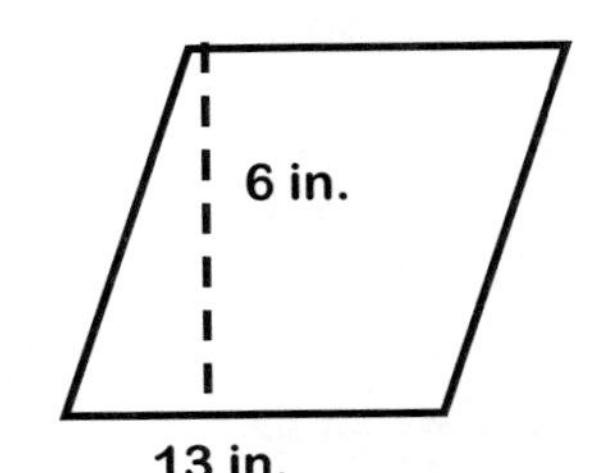

9. Draw intersecting lines.

10. Write the standard number for *fifteen and thirty-six hundredths.*

11. $\frac{6}{10} \div \frac{3}{10} = ?$

12. It is 2:00 now. What time was it 7 hours and 15 minutes ago?

13. $9\frac{1}{5} + 5\frac{1}{4} = ?$

14. Write *0.35* as a percent and as a reduced fraction.

15. A triangle with all sides congruent is a(n) __________ triangle.

16. $4{,}283 \div 9 = ?$

17. Manny can run 660 yards in 3 minutes. At this rate, how many yards can he run in 9 minutes?

18. Draw an acute angle.

19. Which digit is in the thousandths place in *12.0532*?

20. At 8:00 a.m., the temperature was 42°. At 1:00 p.m., it had warmed by 17°, but by 10:00 p.m., it had fallen by 21°. What is the temperature at 10:00 p.m.?

1.	2.	3.	4.
5.	6.	7.	8.
9.	10.	11.	12.
13.	14.	15.	16.
17.	18.	19.	20.

Lesson #5

1. Solve the proportion. $\frac{9}{15} = \frac{a}{10}$
2. $1\frac{1}{2} \times 2\frac{1}{3} = ?$
3. $321 \times 25 = ?$
4. $\frac{8}{9} \times \frac{12}{16} = ?$
5. $90{,}000 - 39{,}816 = ?$
6. Round *34.346* to the nearest hundredth.
7. $27.4 + 6.57 = ?$
8. $\frac{9}{10} \div \frac{3}{5} = ?$
9. Find the surface area of the cube.

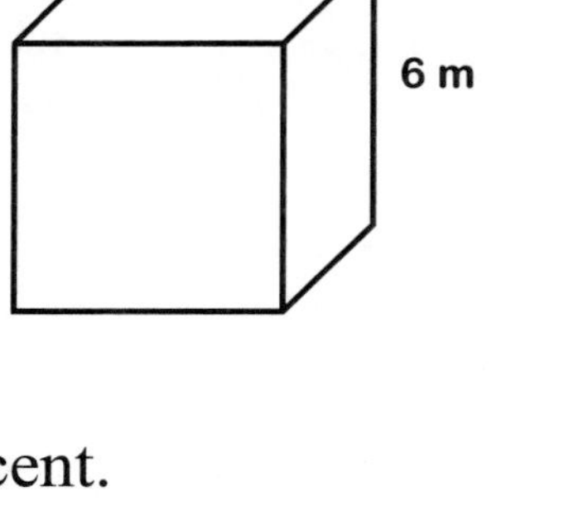

10. $0.035 \times 0.03 = ?$

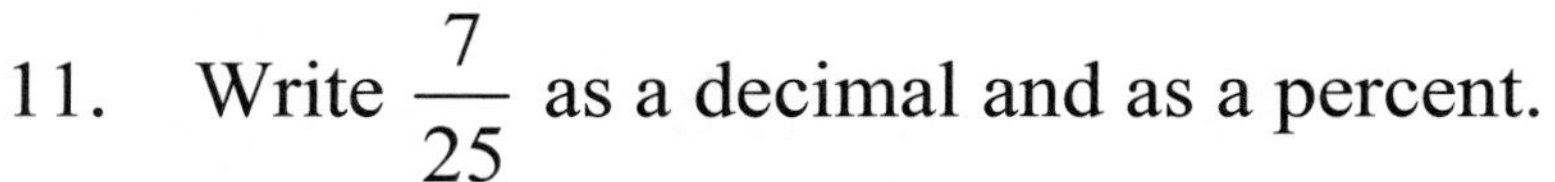

11. Write $\frac{7}{25}$ as a decimal and as a percent.

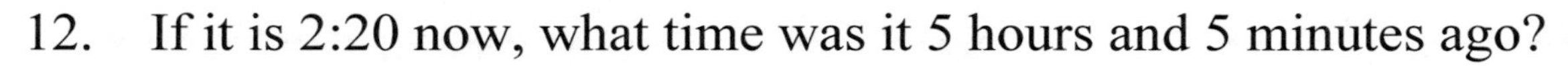

12. If it is 2:20 now, what time was it 5 hours and 5 minutes ago?
13. $\frac{5}{8} \bigcirc \frac{6}{7}$
14. Write *7.1392* using words.
15. How many yards are in 5 miles?
16. $2\frac{1}{3} + 8\frac{2}{5} = ?$
17. Draw a ray.
18. $89{,}364 + 92{,}897 = ?$

19. Make a factor tree for *72*.

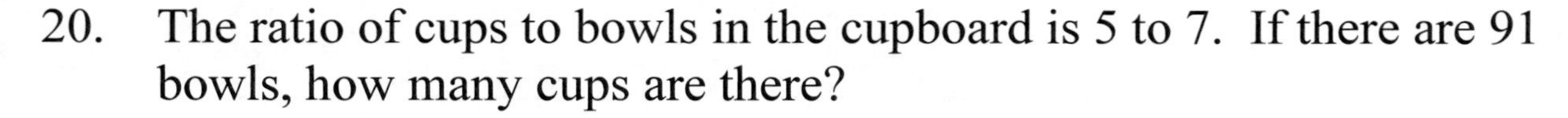

20. The ratio of cups to bowls in the cupboard is 5 to 7. If there are 91 bowls, how many cups are there?

1.	2.	3.	4.
5.	6.	7.	8.
9.	10.	11.	12.
13.	14.	15.	16.
17.	18.	19.	20.

Lesson #6

1. $\frac{5}{7} = \frac{?}{35}$

2. The sum is the answer to a(n) __________ problem.

3. $2\frac{1}{6} \times 3\frac{1}{3} = ?$

4. Draw a line segment.

5. Find the range of 98, 13, 41, 62 and 8.

6. Figures with the same size and shape are ___________.

7. Put $\frac{8}{12}$ in simplest form.

8. $72{,}816 \div 4 = ?$

9. What is the probability of rolling a number greater than 2 on one roll of a die?

10. Round *36,275,813* to the nearest ten million.

11. Write *0.42* as a reduced fraction and as a percent.

12. Which digit is in the tenths place in *25.097*?

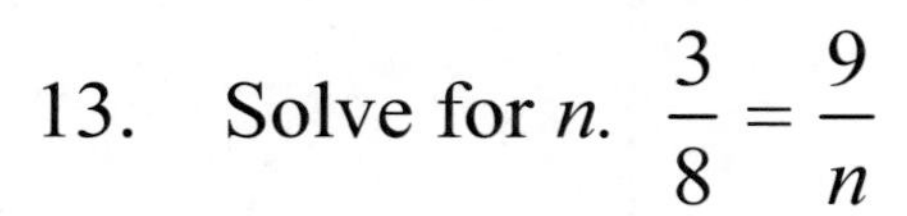

13. Solve for *n*. $\frac{3}{8} = \frac{9}{n}$

14. $5\frac{2}{7} - 1\frac{3}{7} = ?$

15. Find the area of the parallelogram.

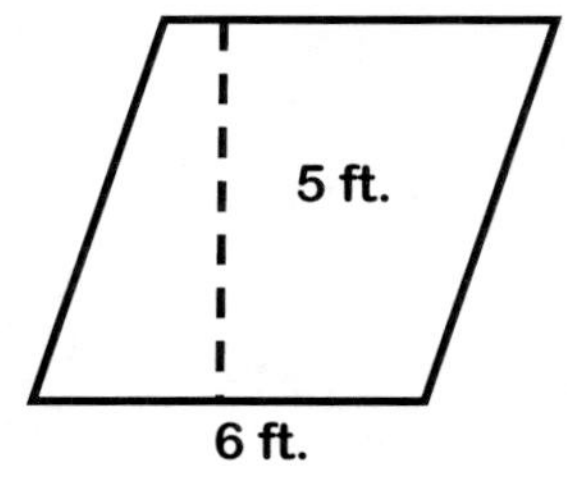

16. $208.4 \div 8 = ?$

17. How many teaspoons are in 8 tablespoons?

18. Draw an obtuse angle.

19. $377{,}519 + 655{,}667 = ?$

20. A decagon has _____________ sides.

1.	2.	3.	4.
5.	6.	7.	8.
9.	10.	11.	12.
13.	14.	15.	16.
17.	18.	19.	20.

Lesson #7

1. $3{,}888 \div 36 = ?$

2. Find the LCM of 5, 8 and 12.

3. On a Celsius thermometer, water boils at what temperature?

4. Write $3\frac{3}{5}$ as an improper fraction.

5. $516{,}677 + 78{,}918 = ?$

6. Find the area of the circle.

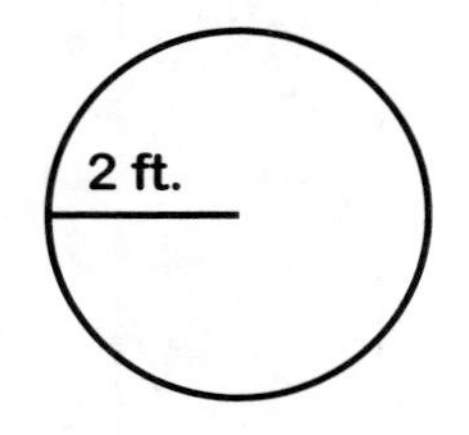

7. $15 - 9\frac{4}{9} = ?$

8. Write *six and four-hundredths* as a decimal number.

9. Closed figures made up of line segments are called _______.

10. $8{,}000 - 2{,}444 = ?$

11. Find the range of 16, 94, 66, 29 and 40.

12. 0.3 ◯ 0.15

13. How many inches are in 3 feet?

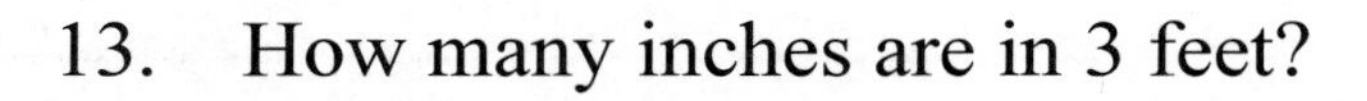

14. Draw 2 similar squares.

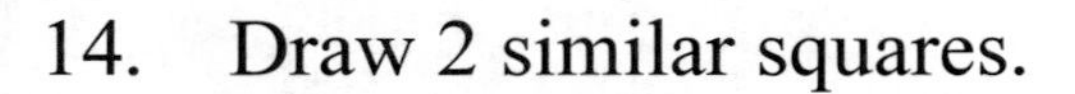

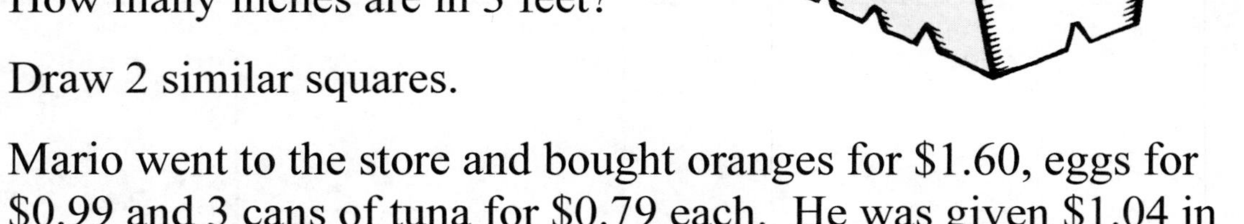

15. Mario went to the store and bought oranges for $1.60, eggs for $0.99 and 3 cans of tuna for $0.79 each. He was given $1.04 in change. How much money did he give the clerk?

16. $0.009 \times 0.06 = ?$

17. What is the name for *the number that occurs most often in a set of numbers*?

18. Find the mean of 320, 294, 265 and 301.

19. The answer to a division problem is called the ________________.

20. $\frac{7}{8} \times \frac{16}{21} = ?$

1.	2.	3.	4.
5.	6.	7.	8.
9.	10.	11.	12.
13.	14.	15.	16.
17.	18.	19.	20.

Lesson #8

1. How many millimeters are in 3 meters?

2. $2\frac{1}{3} \times 1\frac{1}{7} = ?$

3. Write the formula for finding the circumference of a circle.

4. 3,627 – 1,869 = ?

5. Write *57.234* using words.

6. $6 \div \frac{2}{3} = ?$

7. A heptagon has __________ sides.

8. If there are 8 boys and 6 girls in the club, what is the boy to girl ratio?

9. Sue made 3 dozen cupcakes and ate $\frac{1}{6}$ of them. How many are left?

10. 9.2 + 36.9 = ?

11. Numbers that have only 2 factors are called __________ numbers.

12. 325 × 213 = ?

13. Solve for x. $\frac{8}{9} = \frac{x}{27}$

14. 0.56 ÷ 0.4 = ?

15. $7\frac{1}{7} - 3\frac{5}{7} = ?$

16. Round *86.2658* to the nearest hundredth.

17. Find the area of a square if a side measures 12 inches.

18. An angle less than 90° is called a(n) __________ angle.

19. Write the reciprocal of $\frac{5}{6}$.

20. If a truck weighs 4 tons, how many pounds does it weigh?

1.	2.	3.	4.
5.	6.	7.	8.
9.	10.	11.	12.
13.	14.	15.	16.
17.	18.	19.	20.

Lesson #9

1. Jeff has 5 quarters, 2 dimes and 4 pennies. How much money does he have?

2. 80,000 – 46,556 = ?

3. Find the perimeter of the figure.

9 cm

7 cm

4. Find $\frac{4}{5}$ of 25.

5. How many centimeters are in 9 meters?

6. Round *86,716* to the nearest thousand.

7. Write the decimal number *seventeen and seven hundredths*.

8. 547,598 + 761,323 = ?

9. Find the median and mode of 63, 47, 19, 86 and 47.

10. $1.67 \times 0.05 = ?$

11. How many degrees are in a straight angle?

12. Put these decimals in order from least to greatest.
    4.003 4.035 4.305 4.3

13. What will be the time 8 minutes before 2:00?

14. Find the GCF of 12 and 18.

15. Draw perpendicular lines.

16. $4\frac{1}{12} - 3\frac{5}{12} = ?$

17. Write $\frac{3}{20}$ as a decimal and as a percent.

18. $7\frac{1}{3} + 6\frac{1}{8} = ?$

19. Put $\frac{4}{16}$ in simplest form.

20. $13{,}568 \times 8 = ?$

1.	2.	3.	4.
5.	6.	7.	8.
9.	10.	11.	12.
13.	14.	15.	16.
17.	18.	19.	20.

Lesson #10

1. $3{,}007 - 1{,}698 = ?$

2. Use the ruler to find the length of the line segment in inches.

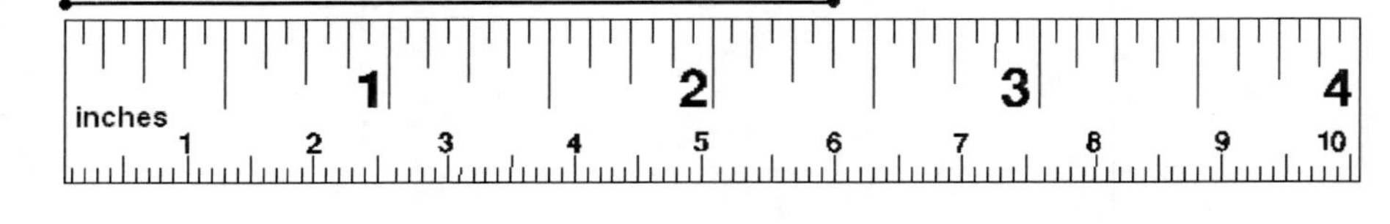

3. $0.6 - 0.459 = ?$

4. $\frac{9}{10} \div \frac{3}{10} = ?$

5. $8.56 + 37.9 = ?$

6. Find the mean of these numbers. 1,121 1,087 1,095

7. What do we call the distance around the outside of a circle?

8. $9\frac{1}{2} - 4\frac{3}{4} = ?$

9. Which is greater, $\frac{1}{5}$ or 35%?

10. Find the GCF of 12 and 16.

11. $98 \times 45 = ?$

12. Solve the proportion. $\frac{6}{10} = \frac{9}{x}$

13. Write $6\frac{2}{3}$ as an improper fraction.

14. How many months are in 3 years?

15. Is the number *23* a prime number or a composite number?

16. Find 40% of 30.

17. Would milliliters or liters be the more reasonable unit for measuring water in a swimming pool?

18. $677 + 999 = ?$

19. How many tons are 14,000 pounds?

20. Draw intersecting lines.

1.	2.	3.	4.
5.	6.	7.	8.
9.	10.	11.	12.
13.	14.	15.	16.
17.	18.	19.	20.

Lesson #11

1. Find the GCF of 15 and 25.

2. Write *36%* as a decimal and as a reduced fraction.

3. $0.525 \div 0.05 = ?$

4. Round *3.982* to the nearest hundredth.

5. $7 - 3\frac{2}{9} = ?$

6. Convert 60 inches to feet.

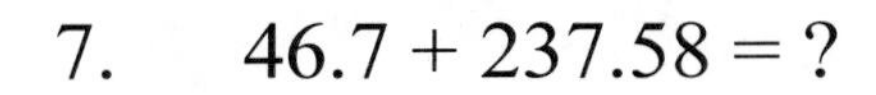

7. $46.7 + 237.58 = ?$

8. $127 \times 216 = ?$

9. Write $\frac{9}{5}$ as a mixed number.

10. How many grams are in a kilogram?

11. What is the probability of rolling a 6 on one roll of a die?

12. $80{,}000 - 38{,}734 = ?$

13. $8\frac{1}{4} + 7\frac{2}{5} = ?$

14. The ratio of cars to trucks at the car show was 9 to 4. If there were 72 cars, how many trucks were at the car show?

15. Find $\frac{3}{5}$ of 45.

16. Find 40% of 70.

17. $\frac{5}{8} \times \frac{16}{25} = ?$

18. Find the mean of these numbers. 2.4 6.3 5.7

19. $0.06 \times 0.03 = ?$

20. If it is 6:20 now, what time will it be in 6 hours and 10 minutes?

1.	2.	3.	4.
5.	6.	7.	8.
9.	10.	11.	12.
13.	14.	15.	16.
17.	18.	19.	20.

Lesson #12

1. Write *12%* as a decimal and as a reduced fraction.

2. Solve for x. $\frac{3}{8} = \frac{9}{x}$

3. $7.5 \div 0.5 = ?$

4. The area of a square is 81 ft^2. What is the measure of each side?

5. $0.6 - 0.2348 = ?$

6. Draw an acute angle.

7. $9\frac{2}{7} + 6\frac{1}{3} = ?$

8. A triangle with 2 sides congruent is what kind of triangle?

9. There are ________ millimeters in a meter.

10. $2\frac{1}{2} \times 1\frac{1}{5} = ?$

11. Find the circumference of the circle.

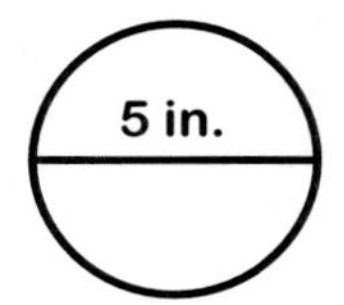

12. Put $\frac{3}{9}$ in simplest form.

13. Write *three and twenty-seven thousandths* as a decimal.

14. Find $\frac{4}{9}$ of 72.

15. Find the perimeter of this polygon.

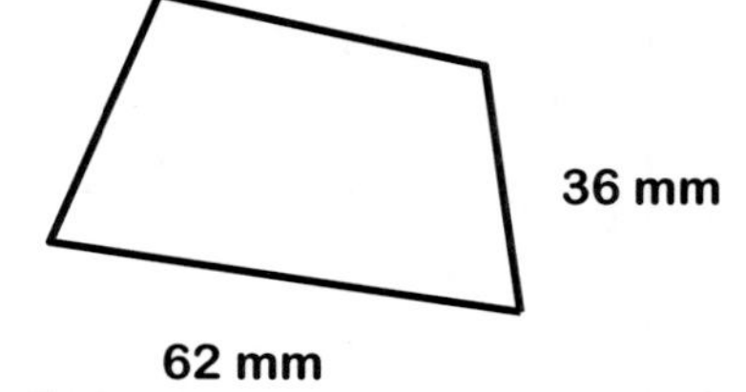

16. $\frac{8}{10} \bigcirc \frac{9}{11}$

17. How many degrees are in a straight angle?

18. The ratio of sparrows to robins in the park was 3 to 8. If there were 104 robins, how many sparrows were in the park?

19. $32{,}107 - 17{,}555 = ?$

20. A number with only 2 factors is called a(n) ____________ number.

1.	2.	3.	4.
5.	6.	7.	8.
9.	10.	11.	12.
13.	14.	15.	16.
17.	18.	19.	20.

Lesson #13

1. What is the estimated product of 387 and 45?

2. Write $\frac{12}{5}$ as a mixed number.

3. Find the area of the rectangle.

11in.

6 in.

4. $70,000 - 35,691 = ?$

5. On the Fahrenheit scale, water boils at ____________.

6. $0.79 \times 0.03 = ?$

7. Write *twelve and two-hundredths* as a decimal.

8. $841,263 + 977,455 = ?$

9. How many teaspoons are in 5 tablespoons?

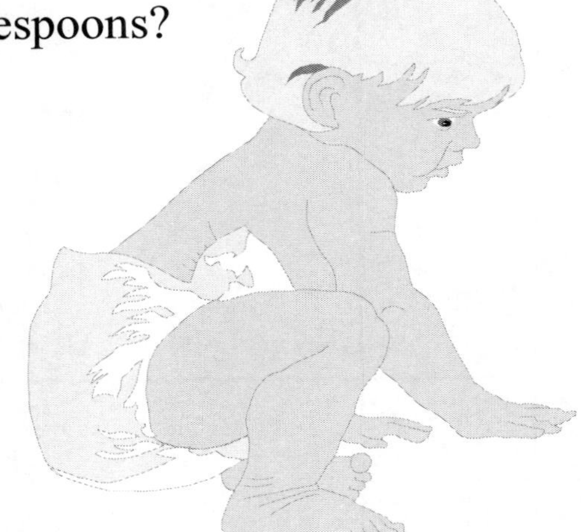

10. $0.4 \bigcirc 0.54$

11. $324 \times 213 = ?$

12. Draw an obtuse angle.

13. $2\frac{1}{2} \div \frac{5}{6} = ?$

14. Round *57.752* to the nearest tenth.

15. Write the formula for finding surface area.

16. A baby weighs 80 ounces. What is the baby's weight in pounds?

17. Is *13* a prime or a composite number?

18. Write *8.1742* using words.

19. $12\frac{1}{6} - 8\frac{5}{6} = ?$

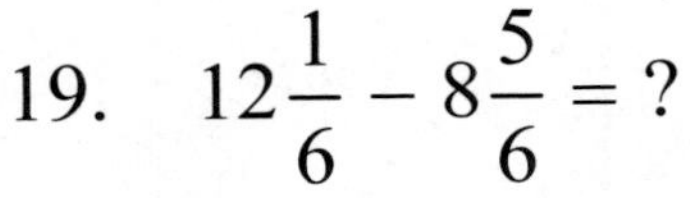

20. How many students earned less than $7?

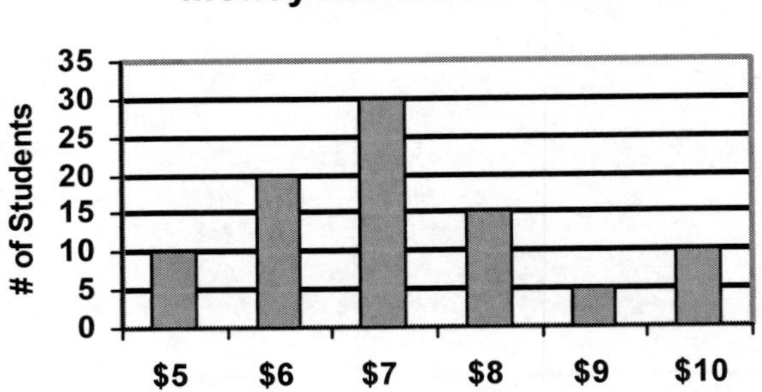

1.	2.	3.	4.
5.	6.	7.	8.
9.	10.	11.	12.
13.	14.	15.	16.
17.	18.	19.	20.

Lesson #14

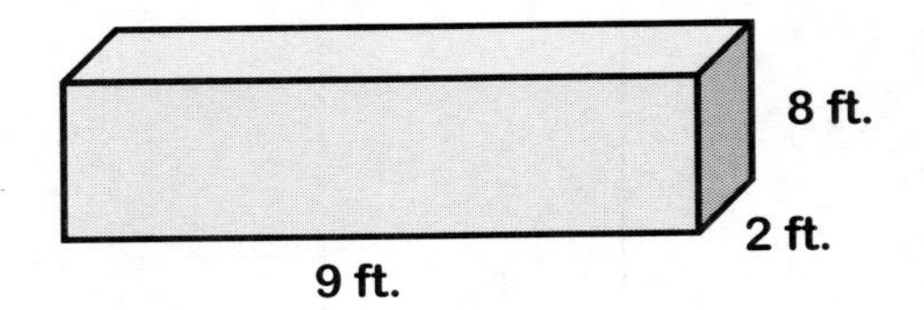

1. Find the volume of the rectangular prism.

2. Find $\frac{3}{4}$ of 24.

3. How many inches are in 4 yards?

4. $8.240 \div 0.4 = ?$

5. Write $\frac{9}{20}$ as a decimal and as a percent.

6. $13.7 - 10.5 = ?$

7. Jack is 5 feet 7 inches tall. What is Jack's height in inches?

8. $0.004 \times 0.07 = ?$

9. The ratio of lions to bears at the zoo was 5 to 7. If there were 91 bears at the zoo, how many lions were there?

10. $6\frac{2}{5} + 9\frac{1}{3} = ?$

11. $865 \div 25 = ?$

12. Write the ratio *5:8* in another way.

13. $\frac{5}{8} \times \frac{12}{15} = ?$

14. $45{,}688 + 973{,}216 = ?$

15. Find the mean of 75, 93, 88 and 60.

16. $\frac{5}{9} = \frac{?}{36}$

17. Draw intersecting lines.

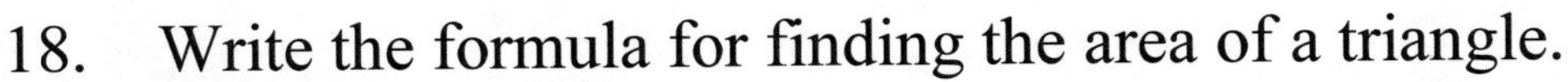

18. Write the formula for finding the area of a triangle.

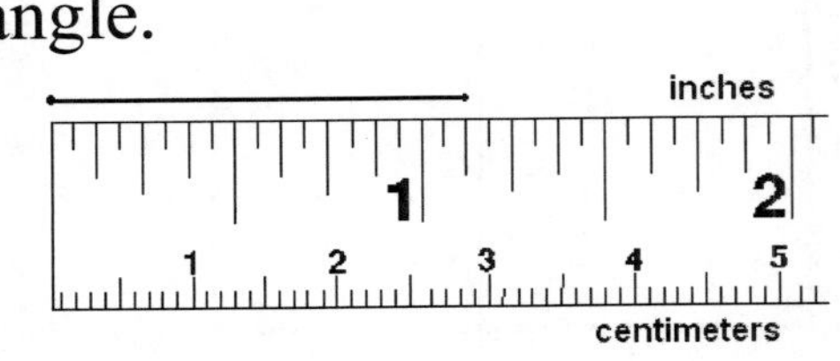

19. Give the length of the line segment in inches.

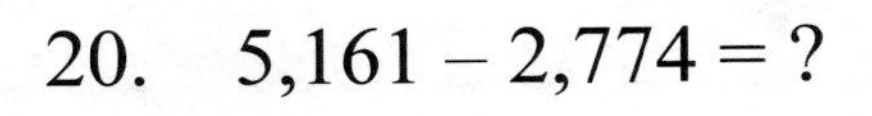

20. $5{,}161 - 2{,}774 = ?$

1.	2.	3.	4.
5.	6.	7.	8.
9.	10.	11.	12.
13.	14.	15.	16.
17.	18.	19.	20.

Lesson #15

1. How many degrees are in a right angle?

2. 633,215 + 845,866 = ?

3. A seven sided polygon is called a(n) ________________.

4. Put $\frac{3}{12}$ in simplest form.

5. 0.07 ◯ 0.083

6. What is the probability of rolling a number greater than 3 on one roll of a die?

7. Find the area of the rectangle.

8. 125 × 216 = ?

9. Find the range of 37, 76, 53 and 25.

10. Which digit is in the hundred thousandths place in *13.46097*?

11. Write $4\frac{2}{7}$ as an improper fraction.

12. 6,000 – 2,323 = ?

13. What is the name of this shape?

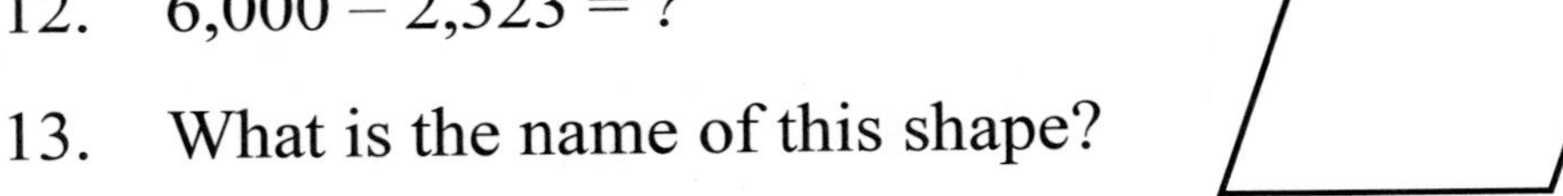

14. 2.7 × 1.3 = ?

15. The answer to a multiplication problem is called the __________.

16. Write the formula for finding the area of a circle.

17. 27.9 – 13.2 = ?

18. Is the dotted line a line of symmetry on the arrow?

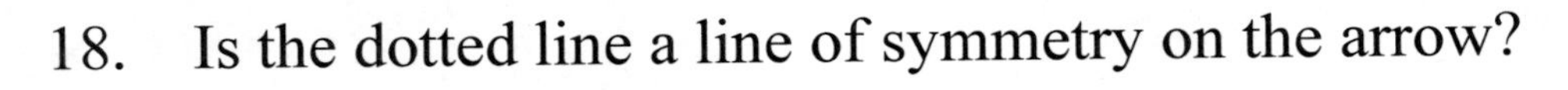

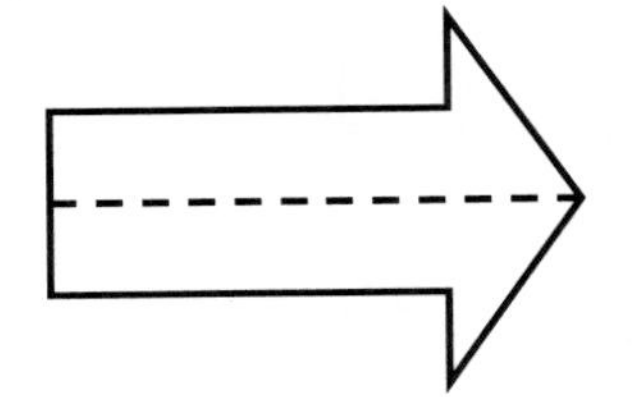

19. Draw perpendicular lines.

20. Delvon had 50 pieces of gum. How many pieces of gum will he have left if he gives $\frac{1}{5}$ of them to his brother and $\frac{2}{5}$ to his friend?

1.	2.	3.	4.
5.	6.	7.	8.
9.	10.	11.	12.
13.	14.	15.	16.
17.	18.	19.	20.

Lesson #16

1. Find the GCF and LCM of 8 and 12.
2. What is the value of x? $\frac{5}{6} = \frac{x}{84}$
3. What are the first 4 prime numbers?
4. Write $\frac{2}{5}$ as a decimal and as a percent.
5. How many millimeters are in a meter?
6. Figures with the same size and shape are ___________.
7. $\frac{5}{6} \times \frac{12}{20} = ?$
8. The answer to a subtraction problem is called the ___________.
9. Write $\frac{11}{5}$ as a mixed number.
10. How many feet are in 2 miles?
11. $7\frac{1}{6} + 5\frac{3}{5} = ?$
12. Put these decimals in order from least to greatest.

    2.075 2.75 2.705 2.5

13. $18.37 + 6.225 = ?$
14. Find the mean of 87, 91, 95, 92 and 85.
15. $\frac{5}{9} \bigcirc \frac{7}{8}$
16. Make a factor tree for *45*.
17. It is 9:30. What time will it be in 10 hours?
18. $8\frac{1}{10} - 4\frac{3}{5} = ?$
19. $0.016 \times 0.004 = ?$
20. How many pounds are in 6 tons?

1.	2.	3.	4.
5.	6.	7.	8.
9.	10.	11.	12.
13.	14.	15.	16.
17.	18.	19.	20.

Lesson #17

1. A class of 30 students ate 9 pizzas. How many pizzas would be needed for 40 students?

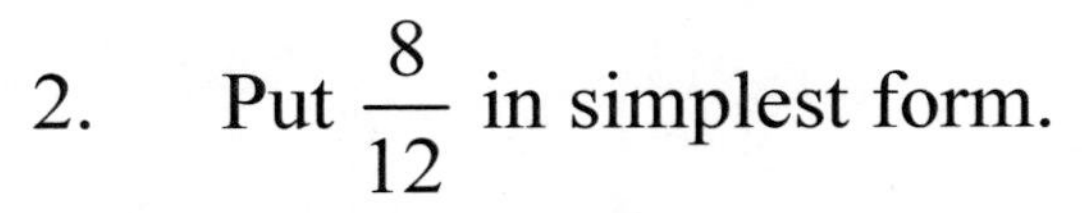

2. Put $\frac{8}{12}$ in simplest form.

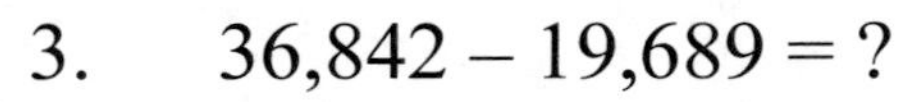

3. $36{,}842 - 19{,}689 = ?$

4. $4\frac{1}{2} \times 2\frac{1}{3} = ?$

5. $2.486 \div 0.02 = ?$

6. Write *25%* as a decimal and as a reduced fraction.

7. $295 \div 14 = ?$

8. Closed figures made up of line segments are _______________.

9. $\frac{9}{10} \bigcirc \frac{3}{4}$

10. Write *2.8675* using words.

11. $\frac{8}{12} \div \frac{3}{4} = ?$

12. Find the area of the parallelogram.

5 ft.

6 ft.

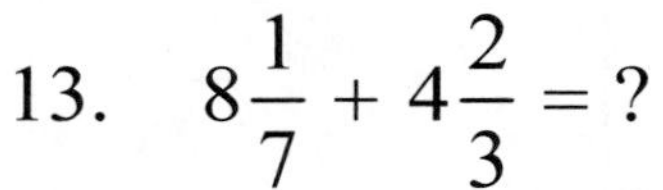

13. $8\frac{1}{7} + 4\frac{2}{3} = ?$

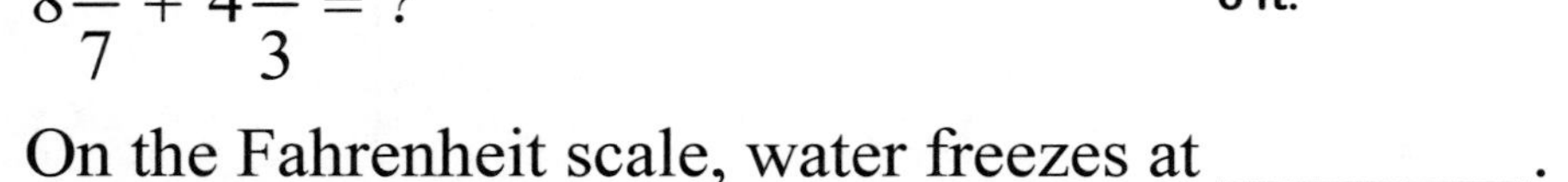

14. On the Fahrenheit scale, water freezes at _________.

15. $47 \times 93 = ?$

16. The answer to a division problem is called the ______________.

17. Solve the proportion. $\frac{3}{7} = \frac{x}{49}$

18. How many yards are in a mile?

19. $8{,}371{,}219 + 7{,}886{,}364 = ?$

20. Draw a ray.

1.	2.	3.	4.
5.	6.	7.	8.
9.	10.	11.	12.
13.	14.	15.	16.
17.	18.	19.	20.

Lesson #18

1. Write *0.15* as a percent and as a reduced fraction.
2. Solve the proportion for x. $\frac{8}{12} = \frac{6}{x}$
3. A nine sided polygon is called a(n) _____________.
4. $10\frac{3}{5} - 8\frac{4}{5} = ?$
5. $486 \times 34 = ?$
6. Find the range of 81, 16, 43, 98 and 11.
7. Numbers that have only 2 factors are _________ numbers.
8. How many millimeters are in 4 meters?
9. What will be the time 90 minutes after 3:30?
10. $90{,}000 - 36{,}642 = ?$
11. Find the area of a square whose sides measure 9 feet.
12. $0.32 \times 0.6 = ?$
13. Three students out of 20 failed the science test. What percent failed the test?
14. What do we call the distance around the outside of a circle?
15. What is 40% of 40?
16. Put these decimals in order from greatest to least.

    1.76		1.607		1.67		1.076
17. A triangle with no sides congruent is called ___________.
18. How many yards are in 4 miles?
19. $0.300 \div 5 = ?$
20. $6\frac{1}{8} + 7\frac{1}{5} = ?$

1.	2.	3.	4.
5.	6.	7.	8.
9.	10.	11.	12.
13.	14.	15.	16.
17.	18.	19.	20.

Lesson #19

1. List the first 5 prime numbers.

2. Draw a line of symmetry through the arrow.

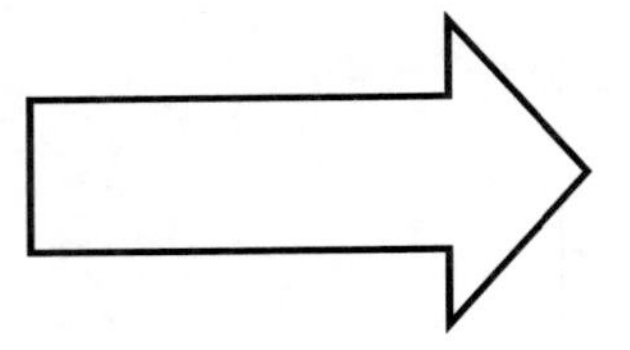

3. $3{,}888 \div 36 = ?$

4. Lauren bought 5 pounds of bacon for $3.75. What is the price for a pound of bacon?

5. Make a factor tree for *64*.

6. Write *sixteen and five-hundredths* as a decimal.

7. $34{,}175 \times 5 = ?$

8. Find the median of 62, 40, 25, 96 and 15.

9. 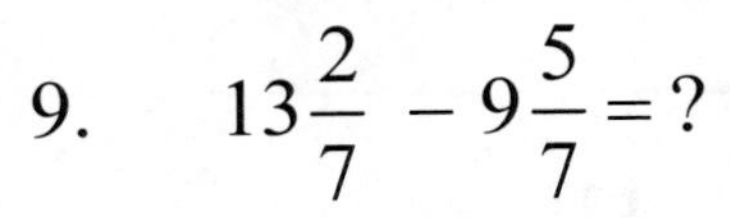 $13\frac{2}{7} - 9\frac{5}{7} = ?$

10. $82{,}730 + 5{,}896 = ?$

11. How many ounces are in 6 pounds?

12. Round *87,342,119* to the nearest million.

13. 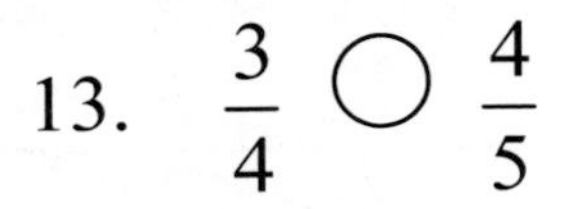 $\frac{3}{4} \bigcirc \frac{4}{5}$

14. How many cups are in 4 pints?

15. 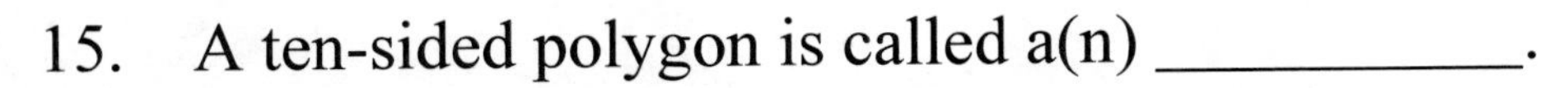 A ten-sided polygon is called a(n) __________.

16. How many feet are the same as 108 inches?

17. Which digit is in the tenths place in *45.8721*?

18. $16.90 - 1.3 = ?$

19. Write $6\frac{1}{3}$ as an improper fraction.

20. What is the coldest time of day?

    Between 8:30 a.m. and noon, how many degrees did the temperature rise?

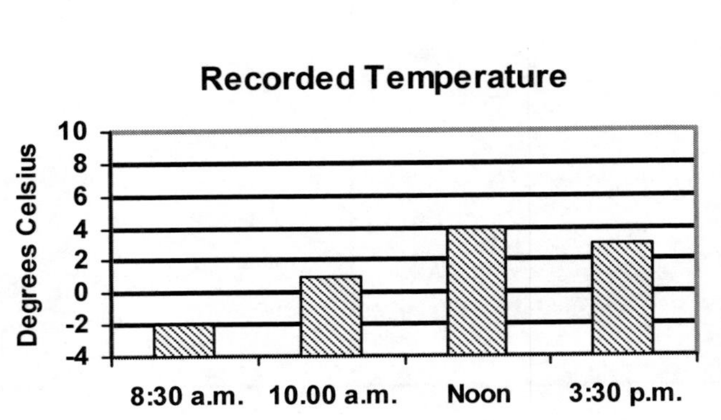

1.	2.	3.	4.
5.	6.	7.	8.
9.	10.	11.	12.
13.	14.	15.	16.
17.	18.	19.	20.

Lesson #20

1. $36.5 + 9.22 + 16.3 = ?$
2. What is $\frac{3}{4}$ of 20?
3. Give the probability of rolling a prime number on one roll of a die.
4. What is the shape of a can of soup?
5. $331 \times 25 = ?$
6. Find the area of the quadrilateral.

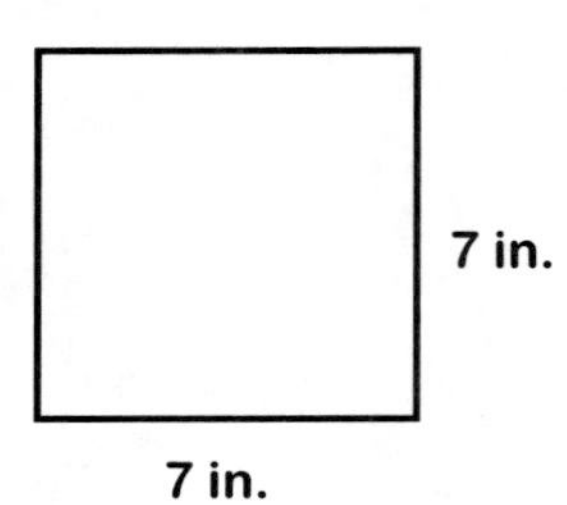

7. $1\frac{1}{2} \times \frac{2}{3} = ?$
8. Find the estimated difference between 8,347 and 3,962.
9. $1.07 \times 0.05 = ?$
10. Write the ratio *6 to 7* in two other ways.
11. Find the mean of 320, 294, 265 and 301.
12. Find the area of a triangle with a base of 12 cm and a height of 4 cm.
13. $5\frac{1}{3} + 9\frac{2}{5} = ?$
14. $5{,}760 \div 8 = ?$
15. How many feet are in 3 miles?
16. Solve the proportion. $\frac{3}{7} = \frac{x}{105}$
17. Write $\frac{7}{20}$ as a decimal and as a percent.
18. How many weeks are 84 days?
19. $\frac{6}{10} \div \frac{3}{10} = ?$
20. Draw an obtuse angle.

1.	2.	3.	4.
5.	6.	7.	8.
9.	10.	11.	12.
13.	14.	15.	16.
17.	18.	19.	20.

Lesson #21

1. Find the GCF of 9 and 12.

2. If it is 7:30 now, what time will it be in 80 minutes?

3. $\frac{9}{10} - \frac{2}{5} = ?$

4. A right angle measures ________ degrees.

5. Round *46.2874* to the nearest hundredth.

6. $146 + 324 + 513 = ?$

7. Write *65%* as a decimal and as a reduced fraction.

8. How many years are 7 centuries?

9. $2.5 \times 1.6 = ?$

10. Find the area of the triangle.

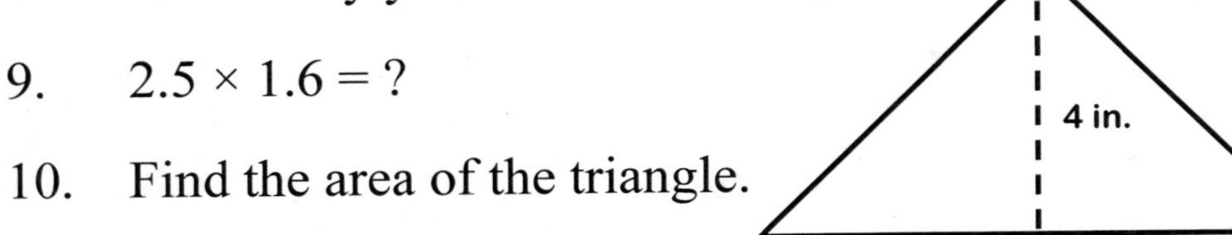

11. $16 - 8\frac{4}{7} = ?$

12. Is *13* a prime or a composite number?

13. $\frac{5}{6} \bigcirc \frac{2}{3}$

14. The answer to a multiplication problem is called the ____________.

15. Put $\frac{20}{25}$ in simplest form.

16. $4.6 - 1.7 = ?$

17. How many pints are in a quart?

18. $1\frac{1}{5} \div 1\frac{1}{2} = ?$

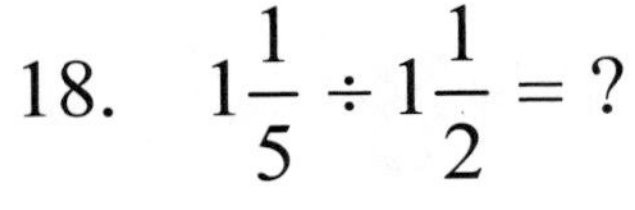

19. Figures with the same size, but different shapes are ___________.

20. A triangle with all of its sides congruent is ___________.

1.	2.	3.	4.
5.	6.	7.	8.
9.	10.	11.	12.
13.	14.	15.	16.
17.	18.	19.	20.

Lesson #22

1. How many millimeters are in 4 meters?
2. $\frac{9}{10} \bigcirc \frac{7}{8}$
3. Find the Least Common Multiple (LCM) of 9 and 12.
4. $0.008 \times 0.004 = ?$
5. Write the reciprocal of $\frac{5}{7}$.
6. List the factors of 18.
7. $6\frac{1}{9} - 4\frac{7}{9} = ?$
8. The ratio of cans to bottles in the pantry is 3 to 4. If there are 64 bottles, how many cans are in the pantry?
9. 48 + _____ = 83
10. Liquid in an eyedropper is best measured in milliliters or in liters?
11. How many quarts are in 15 gallons?
12. $12 \times 12 = ?$
13. Draw a line segment.
14. $1\frac{1}{3} \times \frac{1}{2} = ?$
15. Write *6.21* using words.
16. Half of a circle's diameter is called its __________.
17. $6{,}219 - 3{,}988 = ?$
18. A heptagon has __________ sides.
19. $973 \div 54 = ?$
20. Write $\frac{3}{50}$ as a decimal and as a percent.

1.	2.	3.	4.
5.	6.	7.	8.
9.	10.	11.	12.
13.	14.	15.	16.
17.	18.	19.	20.

Lesson #23

1. $29{,}676 \div 4 = ?$

2. $5\frac{1}{3} + 8\frac{2}{9} = ?$

3. Put these decimals in order from least to greatest.

   45.809 45.908 45.98 45.8

4. Write $8\frac{4}{5}$ as an improper fraction.

5. $\frac{5}{7} \bigcirc \frac{7}{8}$

6. Round *5,887,116* to the nearest hundred thousand.

7. $47 \times 22 = ?$

8. A triangle with 2 sides congruent is called ___________.

9. Which is longer, a meter or a kilometer?

10. $418{,}233 + 26{,}504 = ?$

11. Write *0.55* as a percent and as a reduced fraction.

12. Solve for *x*. $\frac{3}{5} = \frac{x}{75}$

13. $0.9 - 0.465 = ?$

14. A plane weighs 3 tons. What is its weight in pounds?

15. $\frac{8}{9} \times \frac{18}{24} = ?$

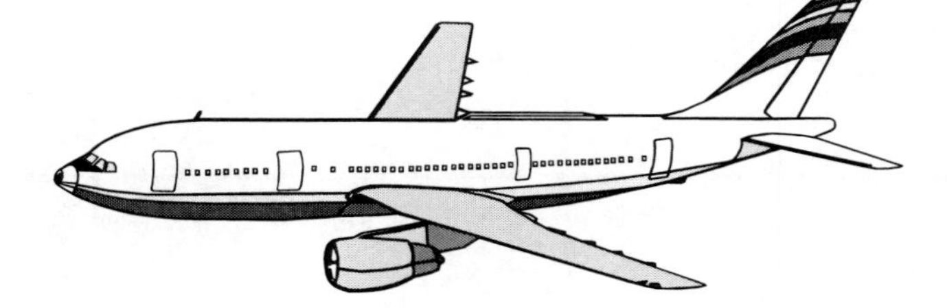

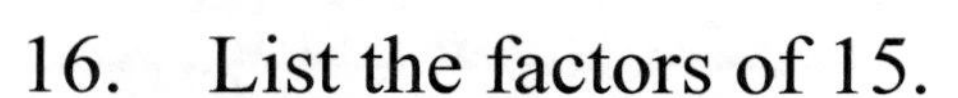

16. List the factors of 15.

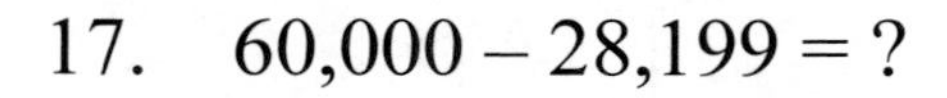

17. $60{,}000 - 28{,}199 = ?$

18. Find the Least Common Multiple (LCM) of 8 and 15.

19. How many teaspoons are in 8 tablespoons?

20. Find the perimeter of a regular hexagon if a side measures 7 inches.

1.	2.	3.	4.
5.	6.	7.	8.
9.	10.	11.	12.
13.	14.	15.	16.
17.	18.	19.	20.

Lesson #24

1. Find the GCF of 15 and 25.

2. The ratio of roses to daisies in the garden was 4 to 7. If there were 64 roses in the garden, how many daisies were there?

3. $0.035 \times 0.06 = ?$

4. $15 - 6\frac{3}{4} = ?$

5. Write *56.7* using words.

6. $6.32 + 15.8 = ?$

7. Draw a straight angle.

8. $\frac{3}{4} \bigcirc \frac{2}{3}$

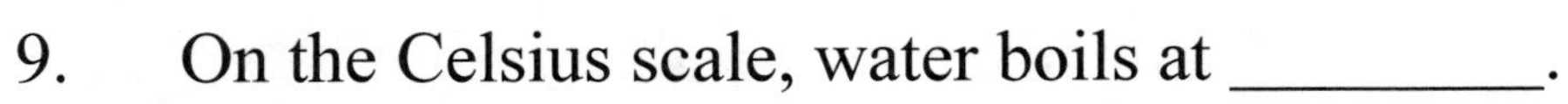

9. On the Celsius scale, water boils at _________.

10. $806 - 233 = ?$

11. How many yards are in 2 miles?

12. $4{,}102 - 1{,}667 = ?$

13. Sean is 5 feet 8 inches tall. What is Sean's height in inches?

14. What will be the time 8 minutes before noon?

15. $72{,}095 \div 45 = ?$

16. Write $5\frac{2}{3}$ as an improper fraction.

17. $7\frac{2}{5} + 3\frac{1}{6} = ?$

18. Put $\frac{3}{15}$ in simplest form.

19. Draw parallel, horizontal lines.

20. What is the probability that the spinner will land on a vowel? On the letter B?

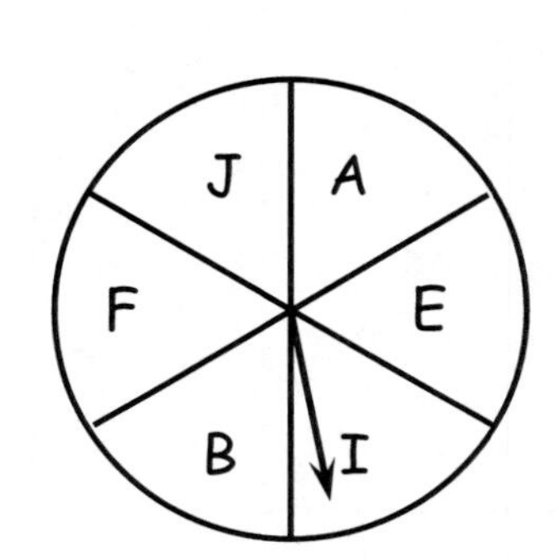

1.	2.	3.	4.
5.	6.	7.	8.
9.	10.	11.	12.
13.	14.	15.	16.
17.	18.	19.	20.

Lesson #25

1. $3.465 \div 5 = ?$

2. Write $\frac{12}{7}$ as a mixed number.

3. Put $\frac{6}{24}$ in simplest form.

4. $\frac{3}{5} \bigcirc \frac{6}{7}$

5. $87.98 + 3.4 = ?$

6. Find the Least Common Multiple (LCM) of 16 and 18.

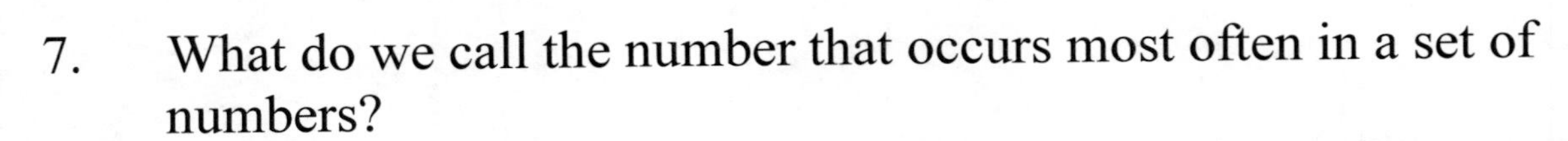

7. What do we call the number that occurs most often in a set of numbers?

8. $503 - 175 = ?$

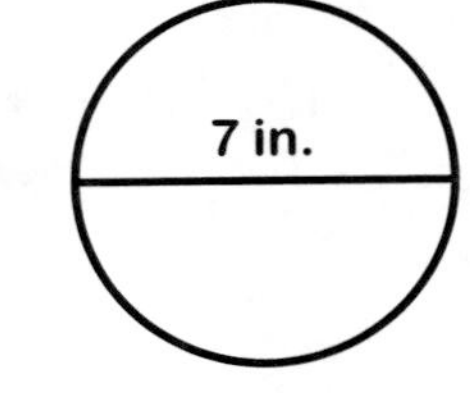

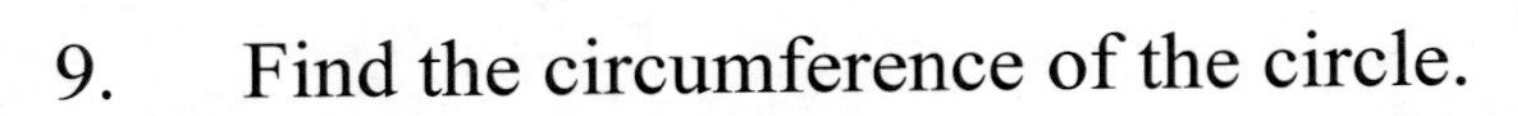

9. Find the circumference of the circle.

10. Find $\frac{5}{6}$ of 24.

11. Find 70% of 50.

12. How many meters are 600 centimeters?

13. Write *75%* as a decimal and as a reduced fraction.

14. Make a factor tree for *24*.

15. Closed figures made up of line segments are called __________.

16. $14\frac{1}{2} - 8\frac{3}{4} = ?$

17. Give the Celsius and Fahrenheit freezing temperatures of water.

18. $\frac{5}{6} \div \frac{2}{3} = ?$

19. How many decades is 80 years?

20. How many grams are in 7 kilograms?

1.	2.	3.	4.
5.	6.	7.	8.
9.	10.	11.	12.
13.	14.	15.	16.
17.	18.	19.	20.

Lesson #26

1. $16.64 \div 0.04 = ?$

2. Find the surface area of this prism.

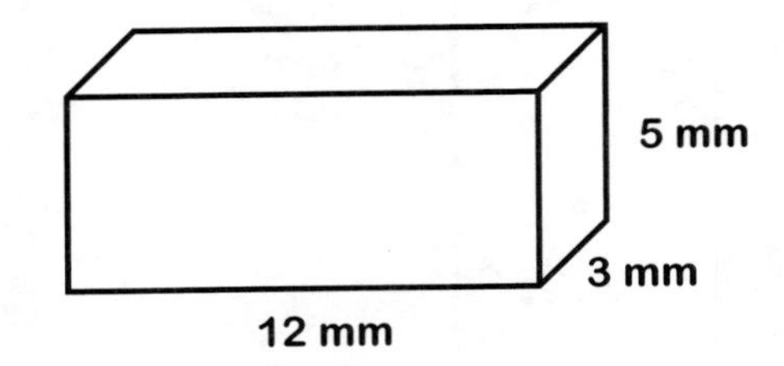

3. $\frac{8}{9} \bigcirc \frac{6}{7}$

4. Find $\frac{3}{7}$ of 49.

5. $5{,}429 + 8{,}552 = ?$

6. Which digit is in the thousands place in *7,815,042*?

7. $20{,}000 - 9{,}335 = ?$

8. Is *28* a prime or a composite number?

9. Draw an acute angle.

10. How many cups are in 3 pints?

11. $551 \times 32 = ?$

12. Write *twenty-five and three-thousandths* as a decimal.

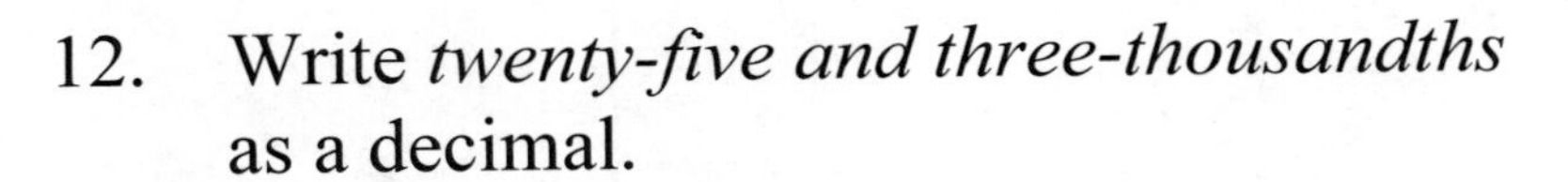

13. Find the Least Common Multiple (LCM) of 8 and 14.

14. $\frac{3}{4} = \frac{?}{20}$

15. The answer to an addition problem is called the ________.

16. Write $\frac{6}{25}$ as a decimal and as a percent.

17. Find the area of a parallelogram if its base is 23 feet and its height is 5 feet.

18. $15 - 6\frac{3}{5} = ?$

19. Are these lines perpendicular?

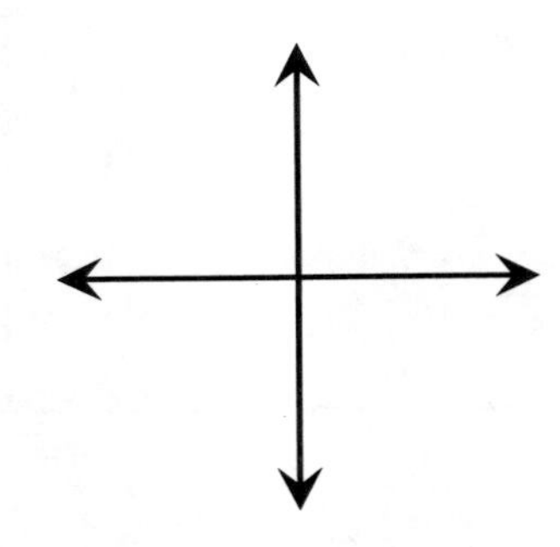

20. Make a factor tree for *32*.

1.	2.	3.	4.
5.	6.	7.	8.
9.	10.	11.	12.
13.	14.	15.	16.
17.	18.	19.	20.

Lesson #27

1. Write *three and seven-thousandths* as a decimal number.

2. Make a factor tree for *45*.

3. Find the perimeter of this regular rectangle.

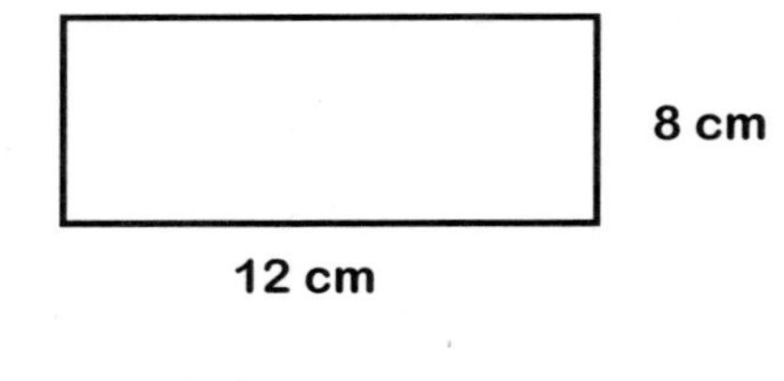

4. $5\frac{4}{9} - 2\frac{7}{9} = ?$

5. Find the mode of these numbers. 14, 25, 55, 14, 62

6. 183,214 + 88,752 = ?

7. How many centimeters are in 5 meters?

8. $0.007 \times 0.04 = ?$

9. If a bird weighs 48 ounces, how many pounds does it weigh?

10. 20,000 – 6,431 = ?

11. A bus weighs 7 tons. What is its weight in pounds?

12. Put $\frac{9}{20}$ as a decimal and as a percent.

13. 0.6 – 0.115 = ?

14. Write the formula for finding the area of a triangle.

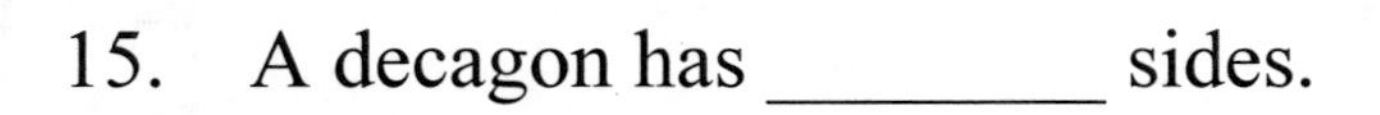

15. A decagon has ________ sides.

16. Find the GCF of 14 and 21.

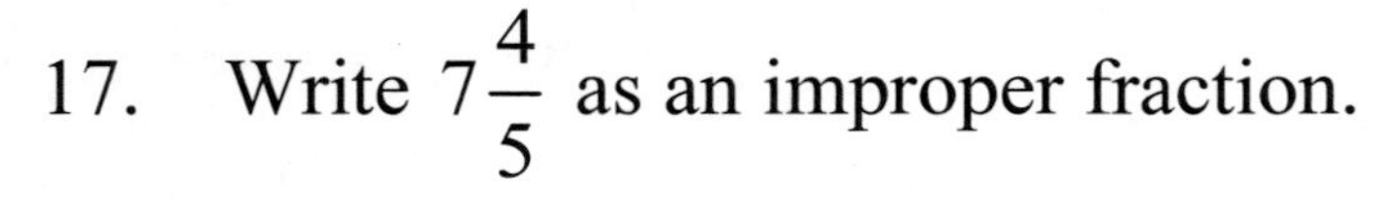

17. Write $7\frac{4}{5}$ as an improper fraction.

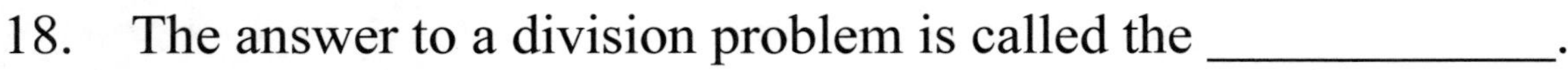

18. The answer to a division problem is called the ___________.

19. $532 \div 21 = ?$

20. Are these shapes congruent or similar?

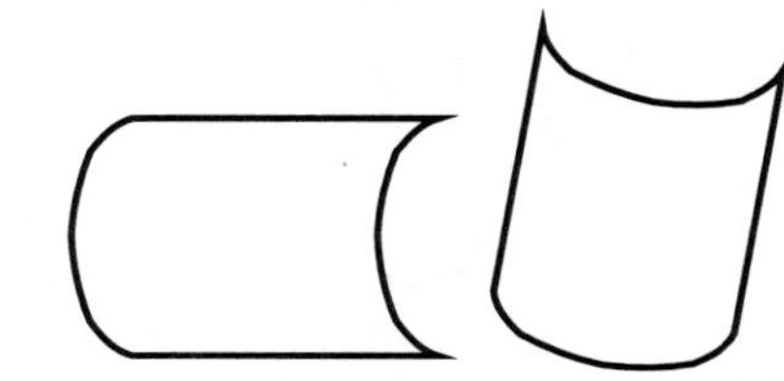

1.	2.	3.	4.
5.	6.	7.	8.
9.	10.	11.	12.
13.	14.	15.	16.
17.	18.	19.	20.

Lesson #28

1. The distance across the middle of a circle is called the _________.

2. If it is 5:30 now, what time was it 9 hours ago?

3. $89 \times 54 = ?$

4. Find $\frac{2}{3}$ of 18.

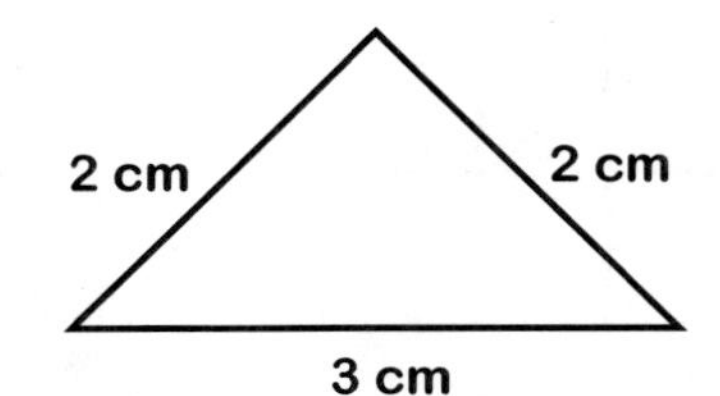

5. Find the perimeter of the triangle.

6. $23{,}513 - 18{,}658 = ?$

7. $\frac{5}{7} \times \frac{14}{25} = ?$

8. Round *23.897* to the nearest tenth.

9. Write the reciprocal of $\frac{3}{4}$.

10. 8.04 ◯ 8.075

11. $5{,}677{,}893 + 3{,}665{,}221 = ?$

12. Write $5\frac{2}{5}$ as an improper fraction.

13. Find the mean of 65, 75 and 85.

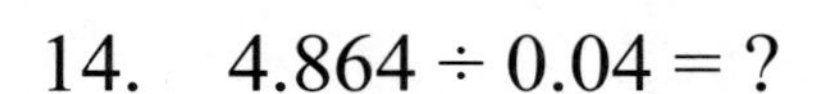

14. $4.864 \div 0.04 = ?$

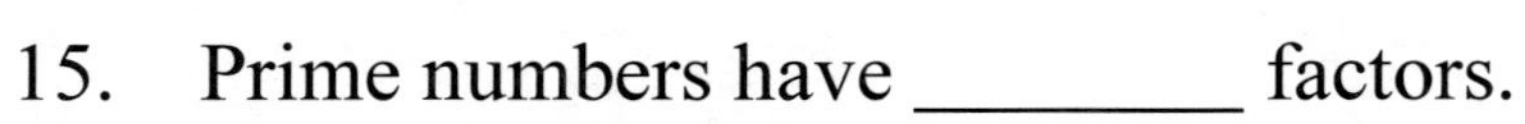

15. Prime numbers have ________ factors.

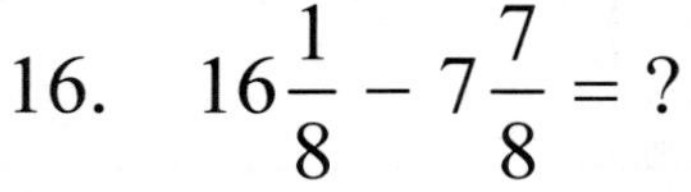

16. $16\frac{1}{8} - 7\frac{7}{8} = ?$

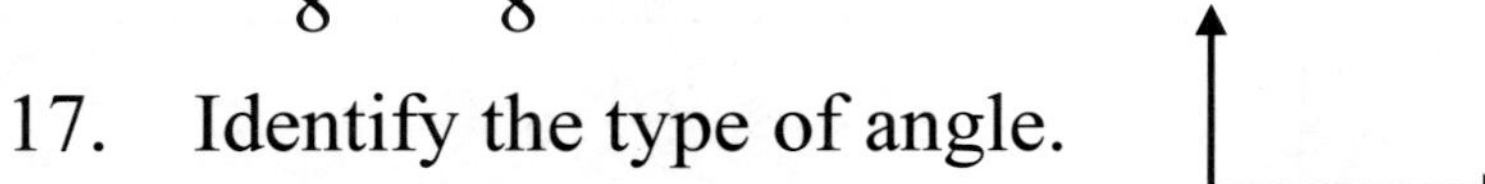

17. Identify the type of angle.

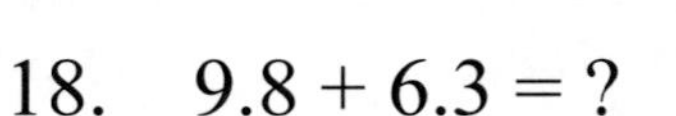

18. $9.8 + 6.3 = ?$

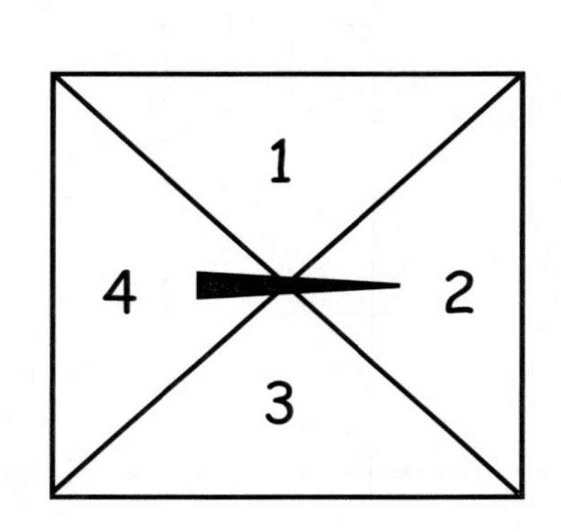

19. How many feet are 5 miles?

20. What is the probability of landing on an even number? On a number greater than 1?

1.	2.	3.	4.
5.	6.	7.	8.
9.	10.	11.	12.
13.	14.	15.	16.
17.	18.	19.	20.

Lesson #29

1. The ratio of spoons to knives in the drawer was 5 to 6. If there were 75 spoons, how many knives were in the drawer?

2. $45{,}671 - 23{,}994 = ?$

3. Write *0.25* as a percent and as a reduced fraction.

4. $3.45 \times 0.09 = ?$

5. Find the volume of this figure.

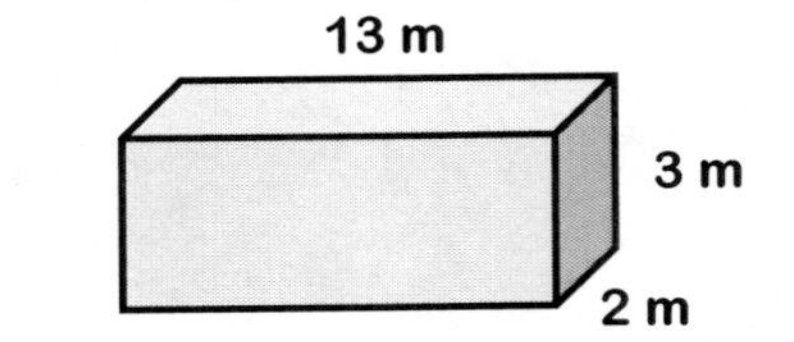

6. Put $\frac{25}{30}$ in simplest form.

7. List the factors of *21*.

8. Give the estimated sum of 89,344 and 22,114.

9. Draw a line segment.

10. $1\frac{1}{5} \times 2\frac{1}{2} = ?$

11. What will be the time 80 minutes after 5:00?

12. $163{,}214 + 79{,}456 = ?$

13. $\frac{7}{10} \bigcirc \frac{8}{9}$

14. Find the GCF and the LCM of 12 and 16.

15. A nine-sided polygon is called a(n) __________.

16. How many yards are in a mile?

17. What is 20% of 60?

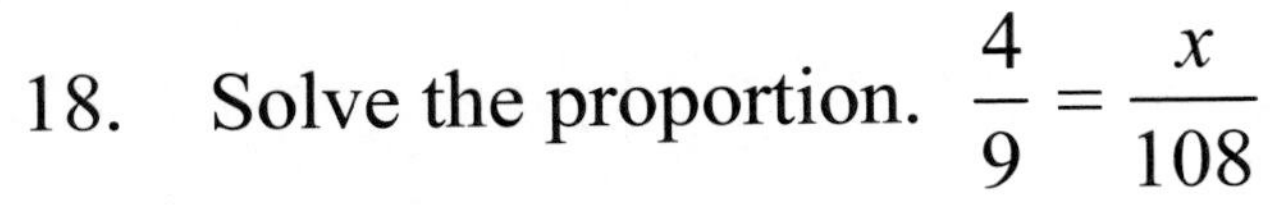

18. Solve the proportion. $\frac{4}{9} = \frac{x}{108}$

19. Put these decimals in order from greatest to least.

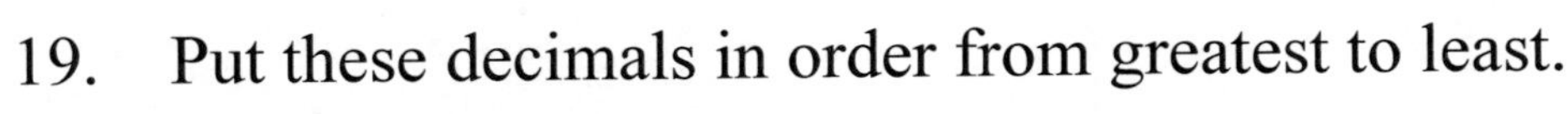

5.668 5.86 5.608 5.8

20. $25{,}678 \times 5 = ?$

1.	2.	3.	4.
5.	6.	7.	8.
9.	10.	11.	12.
13.	14.	15.	16.
17.	18.	19.	20.

Lesson #30

1. A triangle in which all sides are congruent is __________.

2. List the Celsius and the Fahrenheit boiling temperatures of water.

3. Find the mode of 33, 59, 86, 24 and 59.

4. $83 \times 38 = ?$

5. How many inches are in 8 feet?

6. $\frac{6}{7} \times \frac{14}{18} = ?$

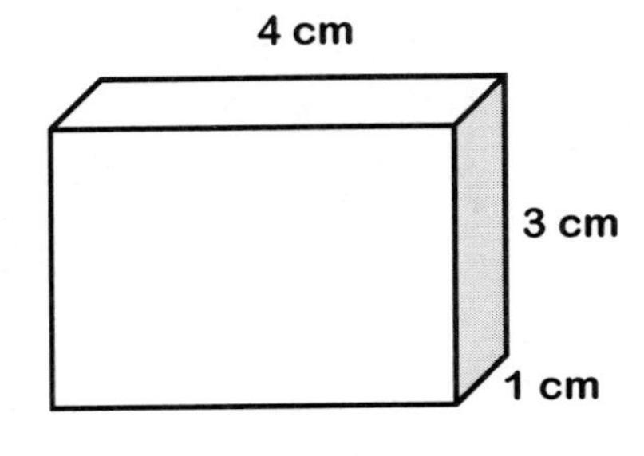

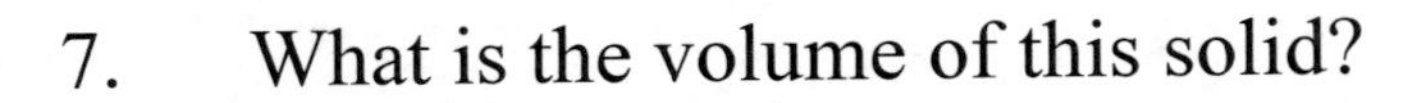

7. What is the volume of this solid?

8. $3{,}963 \div 36 = ?$

9. How many quarts are 6 gallons?

10. Make a factor tree for *56*.

11. $5\frac{2}{9} + 9\frac{1}{3} = ?$

12. Write the formula for finding the area of a parallelogram.

13. If it is 1:25 now, what time will it be in 6 hours and 20 minutes?

14. An angle that measures more than 90° is called __________.

15. $39.63 \div 0.3 = ?$

16. $\frac{8}{9} \div \frac{2}{3} = ?$

17. How many pints are in 8 quarts?

18. $19 - 8\frac{4}{9} = ?$

19. Find the range of 100, 92, 87 and 51.

20. Write *0.85* as a percent and as a reduced fraction.

1.	2.	3.	4.
5.	6.	7.	8.
9.	10.	11.	12.
13.	14.	15.	16.
17.	18.	19.	20.

Intermediate B

Mathematics

Help Pages

Help Pages

Vocabulary

Arithmetic operations

Difference — the result or answer to a subtraction problem. Example: The difference of 5 and 1 is 4.

Product — the result or answer to a multiplication problem. Example: The product of 5 and 3 is 15.

Quotient — the result or answer to a division problem. Example: The quotient of 8 and 2 is 4.

Sum — the result or answer to an addition problem. Example: The sum of 5 and 2 is 7.

Factors and Multiples

Factors — are multiplied together to get a product. Example: 2 and 3 are factors of 6.

Multiples — can be evenly divided by a number. Example: 5, 10, 15 and 20 are multiples of 5.

Composite Number — a number with more than 2 factors.
Example: 10 has factors of 1, 2, 5 and 10. Ten is a composite number.

Prime Number — a number with exactly 2 factors (the number itself & 1). 1 is not prime (it has only 1 factor). Example: 7 has factors of 1 and 7. Seven is a prime number.

Greatest Common Factor (GCF) — the highest factor that 2 numbers have in common.
Example: The factors of 6 are 1, 2, **3** and 6. The factors of 9 are 1, **3** and 9. The GCF of 6 and 9 is 3.

Least Common Multiple (LCM) — the smallest multiple that 2 numbers have in common.
Example: Multiples of 3 are 3, 6, 9, **12**, 15... Multiples of 4 are 4, 8, **12**, 16... The LCM of 3 and 4 is 12.

Prime Factorization — a number written as a product of its prime factors.
Example: 140 can be written as $2 \times 2 \times 5 \times 7$ or $2^2 \times 5 \times 7$. (All of these are prime factors of 140.)

Fractions and Decimals

Improper Fraction — a fraction in which the numerator is larger than the denominator. Example: $\frac{9}{4}$

Mixed Number — the sum of a whole number and a fraction. Example: $5\frac{1}{4}$

Reciprocal — a fraction where the numerator and denominator are interchanged. The product of a fraction and its reciprocal is always 1.
Example: The reciprocal of $\frac{3}{5}$ is $\frac{5}{3}$. $\frac{3}{5} \times \frac{5}{3} = \frac{15}{15} = 1$

Repeating Decimal — a decimal where a number or a series of numbers continues on and on.
Example: 2.33333333, 4.151515151515, 7.125555555. etc.

Geometry

Acute Angle — an angle measuring less than 90°.

Complementary Angles — two angles whose measures add up to 90°.

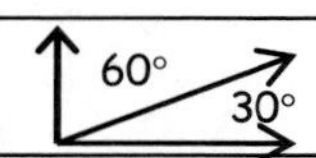

Congruent — figures with the same shape and the same size.

Obtuse Angle — an angle measuring more than 90°.

Right Angle — an angle measuring exactly 90°.

Similar — figures with the same shape but different sizes.

Help Pages

Vocabulary (continued)

Geometry

Straight Angle — an angle measuring exactly 180°.

Supplementary Angles — two angles whose measures add up to 180°.

Surface Area — the sum of the areas of all of the faces of a solid figure.

Geometry — Circles

Circumference — the distance around the outside of a circle.

Diameter — the widest distance across a circle. The diameter always passes through the center.

Radius — the distance from any point on the circle to the center. The radius is half of the diameter.

Geometry — Polygons

Number of Sides	Name	Number of Sides	Name
3	Triangle	7	Heptagon
4	Quadrilateral	8	Octagon
5	Pentagon	9	Nonagon
6	Hexagon	10	Decagon

Geometry — Triangles

Equilateral — a triangle with all 3 sides having the same length.

Isosceles — a triangle with 2 sides having the same length.

Scalene — a triangle with none of the sides having the same length.

Measurement — Relationships

Volume	Distance
3 teaspoons in a tablespoon	36 inches in a yard
2 cups in a pint	1760 yards in a mile
2 pints in a quart	5280 feet in a mile
4 quarts in a gallon	100 centimeters in a meter
Weight	1000 millimeters in a meter
16 ounces in a pound	**Temperature**
2000 pounds in a ton	0° Celsius - Freezing Point
Time	100°Celsius - Boiling Point
10 years in a decade	32°Fahrenheit - Freezing Point
100 years in a century	212°Fahrenheit - Boiling Point

Help Pages

Vocabulary (continued)

Ratio and Proportion

Proportion — a statement that two ratios (or fractions) are equal. Example: $\frac{1}{2} = \frac{3}{6}$

Ratio — a comparison of two numbers by division; a ratio looks like a fraction.

Examples: $\frac{2}{5}$; 2 : 5; 2 to 5; (All of these are pronounced "two to five.")

Percent (%) — the ratio of any number to 100. Example: 14% means 14 out of 100 or $\frac{14}{100}$.

Statistics

Mean — the average of a group of numbers. The mean is found by finding the sum of a group of numbers and then dividing the sum by the number of members in the group.

Example: The mean of 12, 18, 26, 17 and 22 is **19**. $\frac{12+18+26+17+22}{5} = \frac{95}{5} = 19$

Median — the middle value in a group of numbers. The median is found by listing the numbers in order from least to greatest, and finding the one that is in the middle of the list. If there is an even number of members in the group, the median is the average of the two middle numbers.

Example: The median of 14, 17, 24, 11 and 26 is **17**. 11, 14, (17), 24, 26

The median of 77, 93, 85, 95, 70 and 81 is **83**. 70, 77, (81, 85), 93, 95 $\frac{81+85}{2} = 83$

Mode — the number that occurs most often in a group of numbers. The mode is found by counting how many times each number occurs in the list. The number that occurs more than any other is the mode. Some groups of numbers have more than one mode; some have none.

Example: The mode of 77, (93), 85, (93), 77, 81, (93) and 71 is **93**. (93 occurs most often.)

Place Value

Whole Numbers

8,	9	6	3,	2	7	1,	4	0	5
Billions	Hundred Millions	Ten Millions	Millions	Hundred Thousands	Ten Thousands	Thousands	Hundreds	Tens	Ones

The number above is read: eight billion, nine hundred sixty-three million, two hundred seventy-one thousand, four hundred five.

Help Pages

Place Value

Decimal Numbers

1	7	8	.	6	4	0	5	9	2
Hundreds	Tens	Ones	*Decimal Point*	Tenths	Hundredths	Thousandths	Ten-thousandths	Hundred-thousandths	Millionths

The number above is read: one hundred seventy-eight and six hundred forty thousand, five hundred ninety-two millionths.

Solved Examples

Factors & Multiples

The **Prime Factorization** of a number is when a number is written as a product of its prime factors. A factor tree is helpful in finding the prime factors of a number.

Example: Use a factor tree to find the prime factors of 45.

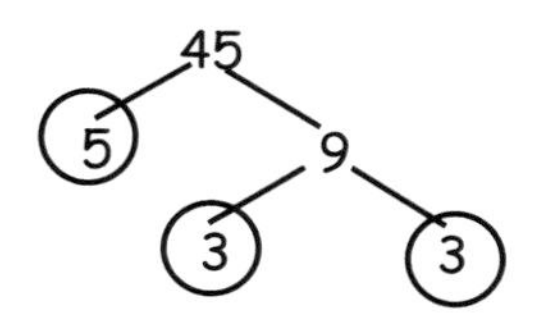

1. Find any 2 factors of 45 (5 & 9).
2. If a factor is prime, circle it. If a factor is not prime, find 2 factors of it.
3. Continue until all factors are prime.
4. In the final answer, the prime factors are listed in order, least to greatest, using exponents when needed.

The prime factorization of 45 is $3 \times 3 \times 5$ or $3^2 \times 5$.

The **Greatest Common Factor (GCF)** is the largest factor that 2 numbers have in common.

Example: Find the Greatest Common Factor of 32 and 40.

The factors of 32 are 1, 2, 4, (8), 16, 32.

The factors of 40 are 1, 2, 4, 5, (8), 10, 20, 40.

1. First list the factors of each number.
2. Find the largest number that is in both lists.

The GCF of 32 and 40 is **8**.

Help Pages

Solved Examples

Factors & Multiples (continued)

The **Least Common Multiple (LCM)** is the smallest multiple that two numbers have in common. The prime factors of the numbers can be useful in finding the LCM.

Example: Find the Least Common Multiple of 16 and 24.

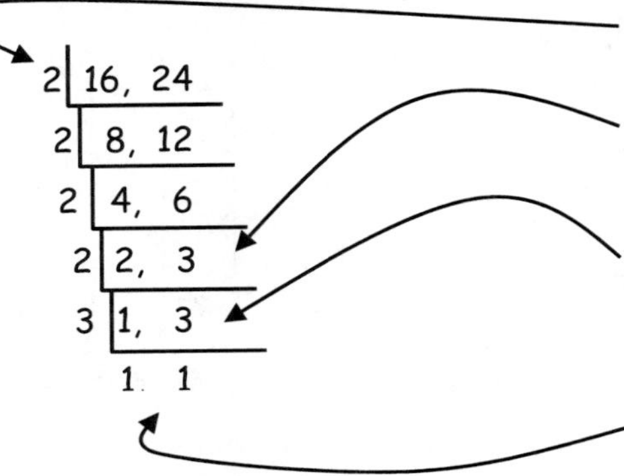

1. If any of the numbers are even, factor out a 2.
2. Continue factoring out 2 until all numbers left are odd.
3. If the prime number cannot be divided evenly into the number, simply bring the number down.
4. Once you are left with all 1's at the bottom, you're finished!
5. Multiply all of the prime numbers (on the left side of the bracket) together to find the Least Common Multiple.

The LCM is $2 \times 2 \times 2 \times 2 \times 3$ or **48**.

Fractions

Changing from one form to another...

Example: **Change the improper fraction, $\frac{5}{2}$, to a mixed number.**

$\frac{5}{2}$ (five halves) means $5 \div 2$.

So, $\frac{5}{2}$ is equal to 2 wholes and 1 half or $2\frac{1}{2}$.

$$\begin{array}{r} 2 \text{ wholes} \\ 2\overline{)5} \\ -4 \\ \hline 1 \text{ half} \end{array}$$

Changing from one form to another...

Example: **Change the mixed number, $7\frac{1}{4}$, to an improper fraction.**

1. You're going to make a new fraction. To find the numerator of the new fraction, multiply the whole number by the denominator, and add the numerator.
2. Keep the same denominator in your new fraction as you had in the mixed number.

$7\frac{1}{4}$ $7 \times 4 = 28$. $28 + 1 =$ **29**.

The new numerator is 29.

Keep the same denominator, 4.

The new fraction is $\frac{29}{4}$.

$7\frac{1}{4}$ is equal to $\frac{29}{4}$.

Help Pages

Solved Examples

Fractions (continued)

Equivalent Fractions are 2 fractions that are equal to each other. Usually you will be finding a missing numerator or denominator.

Example: Find a fraction that is equivalent to $\frac{4}{5}$ and has a denominator of 35.

$$\frac{4}{5} \overset{\times 7}{=} \frac{?}{35}$$

1. Ask yourself, "What did I do to 5 to get 35?" (Multiply by 7.)
2. Whatever you did in the denominator, you also must do in the numerator. 4 x 7 = 28. The missing numerator is 28.

So, $\frac{4}{5}$ is equivalent to $\frac{28}{35}$.

Example: Find a fraction that is equivalent to $\frac{4}{5}$ and has a numerator of 24.

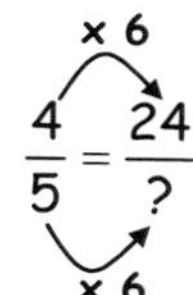

1. Ask yourself, "What did I do to 4 to get 24?" (Multiply by 6.)
2. Whatever you did in the numerator, you also must do in the denominator. 5 x 6 = 30. The missing denominator is 30.

So, $\frac{4}{5}$ is equivalent to $\frac{24}{30}$.

Comparing Fractions means looking at 2 or more fractions and determining if they are equal, if one is greater than (>) the other or if one is less than (<) the other. A simple way to compare fractions is by cross-multiplying, using the steps below.

Examples: Compare these fractions. Use the correct symbol. $\frac{8}{9} \bigcirc \frac{3}{4}$ $\frac{7}{9} \bigcirc \frac{6}{7}$

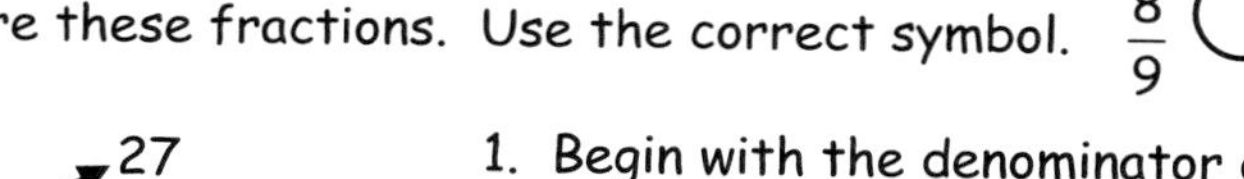

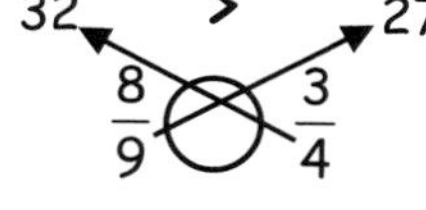

49 < 54 ($\frac{7}{9}$, $\frac{6}{7}$)

1. Begin with the denominator on the left and multiply by the opposite numerator. Put the answer (product) above the side where you ended. (9 x 3 = 27)
2. Cross-multiply the other denominator and numerator and put that answer above where you ended.
3. Compare the two answers and insert the correct symbol.

HINT: **Always** multiply diagonally **upwards**!

So, $\frac{8}{9} > \frac{3}{4}$ and $\frac{7}{9} < \frac{6}{7}$.

To **add (or subtract) fractions with the same denominator**, simply add (or subtract) the numerators, keeping the same denominator.

Examples: $\frac{3}{5} + \frac{1}{5} = \frac{4}{5}$

$\frac{8}{9} - \frac{1}{9} = \frac{7}{9}$

When **adding mixed numbers**, follow a similar process as you used with fractions. If the sum is an improper fraction, make sure to simplify it.

Example: $1\frac{2}{5} + 1\frac{4}{5} = 2\frac{6}{5}$

$2\frac{6}{5}$ is improper. $\frac{6}{5}$ can be rewritten as $1\frac{1}{5}$.

So, $2\frac{6}{5}$ is $2 + 1\frac{1}{5} = 3\frac{1}{5}$.

Help Pages

Solved Examples

Fractions (continued)

When **adding fractions that have different denominators**, you first need to change the fractions so they have a common denominator. Then, you can add them.

Finding the **Least Common Denominator (LCD):**

The LCD of the fractions is the same as the Least Common Multiple of the denominators. Sometimes, the LCD will be the product of the denominators.

Example: Find the sum of $\frac{3}{8}$ and $\frac{1}{12}$.

$$\frac{3}{8} = \frac{9}{24}$$
$$+\frac{1}{12} = \frac{2}{24}$$
$$\frac{11}{24}$$

1. First, find the LCM of 8 and 12.
2. The LCM of 8 and 12 is 24. This is also the LCD of these 2 fractions.
3. Find an equivalent fraction for each that has a denominator of 24.
4. When they have a common denominator, the fractions can be added.

2 | 8,12
2 | 4, 6
2 | 2,3
3 | 1,3
1, 1

$2 \times 2 \times 2 \times 3 = 24$

The LCM is 24.

Example: Add $\frac{1}{4}$ and $\frac{1}{5}$.

$$\frac{1}{4} = \frac{5}{20}$$
$$+\frac{1}{5} = \frac{4}{20}$$
$$\frac{9}{20}$$

$4 \times 5 = 20$ The LCM is 20.

When **adding mixed numbers with unlike denominators**, follow a process similar to the one you used with fractions (above). Make sure to put your answer in simplest form.

Example: Find the sum of $6\frac{3}{7}$ and $5\frac{2}{3}$.

$$6\frac{3}{7} = 6\frac{9}{21}$$
$$+5\frac{2}{3} = 5\frac{14}{21}$$
$$11\frac{23}{21}$$

(improper)

$$\frac{23}{21} = 1\frac{2}{21} + 11 = 12\frac{2}{21}$$

1. Find the LCD.
2. Make new equivalent fractions using the LCD.
3. Add the whole numbers, then add the fractions.
4. Make sure your answer is in simplest form.

Help Pages

Solved Examples

Fractions (continued)

When **subtracting numbers with unlike denominators**, follow a process similar to the one you used when adding fractions. Make sure to put your answer in simplest form.

Examples: Find the difference of $\frac{3}{4}$ and $\frac{2}{5}$.

$$\begin{array}{r} \frac{3}{4} = \frac{15}{20} \\ -\frac{2}{5} = \frac{8}{20} \\ \hline \frac{7}{20} \end{array}$$

1. Find the LCD just as you did when adding fractions.
2. Make new equivalent fractions using the LCD.
3. Subtract the numerators and keep the common denominator.
4. Make sure your answer is in simplest form.

Subtract $\frac{1}{16}$ from $\frac{3}{8}$.

$$\begin{array}{r} \frac{3}{8} = \frac{6}{16} \\ -\frac{1}{16} = \frac{1}{16} \\ \hline \frac{5}{16} \end{array}$$

When **subtracting mixed numbers with unlike denominators**, follow a process similar to the one you used when adding mixed numbers. Make sure to put your answer in simplest form.

Example: Subtract $4\frac{2}{5}$ from $8\frac{9}{10}$.

1. Find the LCD.
2. Make new equivalent fractions using the LCD.
3. Subtract and simplify your answer.

$$\begin{array}{r} 8\frac{9}{10} = 8\frac{9}{10} \\ -4\frac{2}{5} = 4\frac{4}{10} \\ \hline 4\frac{5}{10} = 4\frac{1}{2} \end{array}$$

Sometimes when subtracting mixed numbers, you may need to regroup. If the numerator of the top fraction is smaller than the numerator of the bottom fraction, you must borrow from your whole number.

Example: Subtract $5\frac{5}{6}$ from $9\frac{1}{4}$.

$$9\frac{1}{4} = 9\frac{3}{12} = 8\frac{12}{12} + \frac{3}{12} = 8\frac{15}{12}$$

$$-5\frac{5}{6} = 5\frac{10}{12} = \quad -5\frac{10}{12}$$

$$3\frac{5}{12}$$

1. Find the LCD.
2. Make new equivalent fractions using the LCD.
3. Because you can't subtract 10 from 3, you need to borrow from the whole number.
4. Rename the whole number as a mixed number using the common denominator.
5. Add the 2 fractions to get an improper fraction.
6. Subtract the whole numbers and the fractions and simplify your answer.

More examples:

$$8\frac{1}{2} = 8\frac{2}{4} = 7\frac{4}{4} + \frac{2}{4} = 7\frac{6}{4}$$

$$-4\frac{3}{4} = 4\frac{3}{4} = \quad -4\frac{3}{4}$$

$$3\frac{3}{4}$$

$$10\frac{1}{5} = 10\frac{4}{20} = 9\frac{20}{20} + \frac{4}{20} = 9\frac{24}{20}$$

$$-6\frac{3}{4} = 6\frac{15}{20} = \quad -6\frac{15}{20}$$

$$3\frac{9}{20}$$

Help Pages

Solved Examples

Fractions (continued)

To **multiply fractions**, simply multiply the numerators together to get the numerator of the product. Then multiply the denominators together to get the denominator of the product. Make sure your answer is in simplest form.

Examples: Multiply $\frac{3}{5}$ by $\frac{2}{3}$.

$$\frac{3}{5} \times \frac{2}{3} = \frac{6}{15} = \frac{2}{5}$$

1. Multiply the numerators.
2. Multiply the denominators.
3. Simplify your answer.

Multiply $\frac{5}{8}$ by $\frac{4}{5}$.

$$\frac{5}{8} \times \frac{4}{5} = \frac{20}{40} = \frac{1}{2}$$

Sometimes you can use canceling when multiplying fractions. Let's look at the examples again.

$$\frac{\cancel{3}^{1}}{5} \times \frac{2}{\cancel{3}_{1}} = \frac{2}{5}$$

The 3's have a common factor: 3. Divide both of them by 3. Since $3 \div 3 = 1$, we cross out the 3's and write 1 in its place.

Now, multiply the fractions. In the numerator, $1 \times 2 = 2$. In the denominator, $5 \times 1 = 5$.

The answer is $\frac{2}{5}$.

1. Are there any numbers in the numerator and the denominator that have common factors?
2. If so, cross out the numbers, divide both by that factor, and write the quotient.
3. Then, multiply the fractions as described above, using the quotients instead of the original numbers.

$$\frac{\cancel{5}^{1}}{{}_{2}\cancel{8}} \times \frac{\cancel{4}^{1}}{\cancel{5}_{1}} = \frac{1}{2}$$

As in the other example, the 5's can be cancelled. But here, the 4 and the 8 also have a common factor: 4. $8 \div 4 = 2$ and $4 \div 4 = 1$. After canceling both of these, you can multiply the fractions.

REMEMBER: You can cancel up and down or diagonally, but NEVER sideways!

When **multiplying mixed numbers**, you must first change them into improper fractions.

Examples: Multiply $2\frac{1}{4}$ by $3\frac{1}{9}$.

$$2\frac{1}{4} \times 3\frac{1}{9} =$$

$$\frac{\cancel{9}^{1}}{{}_{1}\cancel{4}} \times \frac{\cancel{28}^{7}}{\cancel{9}_{1}} = \frac{7}{1} = 7$$

1. Change each mixed number to an improper fraction.
2. Cancel wherever you can.
3. Multiply the fractions.
4. Put your answer in simplest form.

Multiply $3\frac{1}{8}$ by 4.

$$3\frac{1}{8} \times 4 =$$

$$\frac{25}{{}_{2}\cancel{8}} \times \frac{\cancel{4}^{1}}{1} = \frac{25}{2} = 12\frac{1}{2}$$

To **divide fractions**, you must take the reciprocal of the 2nd fraction, and then multiply that reciprocal by the 1st fraction. Don't forget to simplify your answer.

Examples: Divide $\frac{1}{2}$ by $\frac{7}{12}$.

$$\frac{1}{2} \div \frac{7}{12} =$$

$$\frac{1}{{}_{1}\cancel{2}} \times \frac{\cancel{12}^{6}}{7} = \frac{6}{7}$$

1. Keep the 1st fraction as it is.
2. Write the reciprocal of the 2nd fraction.
3. Change the sign to multiplication.
4. Cancel if you can and multiply.
5. Simplify your answer.

Divide $\frac{7}{8}$ by $\frac{3}{4}$.

$$\frac{7}{8} \div \frac{3}{4} =$$

$$\frac{7}{{}_{2}\cancel{8}} \times \frac{\cancel{4}^{1}}{3} = \frac{7}{6} = 1\frac{1}{6}$$

Help Pages

Solved Examples

Fractions (continued)

When **dividing mixed numbers**, you must first change them into improper fractions.

Example: Divide $1\frac{1}{4}$ by $3\frac{1}{2}$.

$$1\frac{1}{4} \div 3\frac{1}{2} =$$

$$\frac{5}{4} \div \frac{7}{2} =$$

$$\frac{5}{\cancel{4}_{2}} \times \frac{\cancel{2}^{1}}{7} = \frac{5}{14}$$

1. Change each mixed number to an improper fraction.
2. Keep the 1st fraction as it is.
3. Write the reciprocal of the 2nd fraction.
4. Change the sign to multiplication.
5. Cancel if you can and multiply.
6. Simplify your answer.

Decimals

When **comparing decimals**, the purpose is to decide which has the smaller or larger value. They are sometimes compared by placing them in order from least to greatest or from greatest to least. Another way to compare is to use the symbols for "less than" (<), "greater than" (>) or "equal to" (=).

Example: Order these numbers from least to greatest. 0.561 0.506 0.165

1. Write the numbers in a column, lining up the decimal points.
2. Write zeroes, if necessary, so all have the same number of digits.
3. Begin on the left and compare the digits.

0.561
0.506
0.165

Since they all have 3 digits, zeroes don't need to be added. Beginning on the left, the five's are equal, but the one is less, so 0.165 is the smallest. Then, look at the next digit. The zero is less than the six, so 0.506 is next smallest.

So, in order from least to greatest:

0.165, 0.506, 0.561

Example: Place these numbers in order from greatest to least. 0.44 0.463 0.045

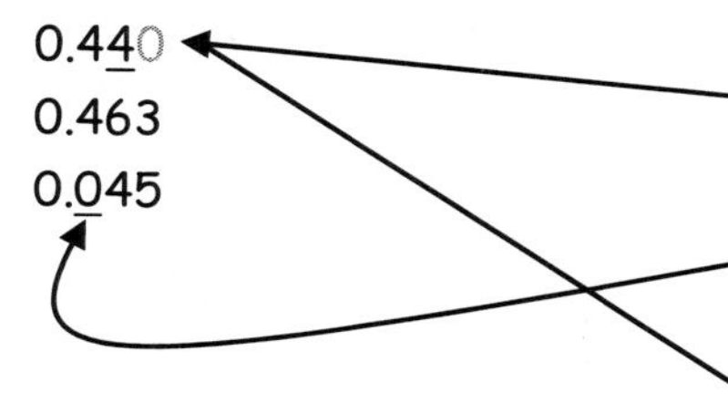

After lining up the numbers, we must add a zero to 0.44 to make them all have the same number of digits.

Beginning on the left, the zero is smaller than the four's, so 0.045 is the smallest.

Look at the next digit. The four is smaller than the six, so 0.440 is the next smallest.

In order from greatest to least: 0.463, 0.440, 0.045

Help Pages

Solved Examples

Decimals (continued)

When **rounding decimals**, they are being approximated. This means the decimal ends at a certain place value and is either rounded up (if it's closer to the next higher number) or kept the same (if it's closer to the next lower number). It might be helpful to look at the decimal place-value chart on p. 67.

Example: Round 0.574 to the tenths place.

There is a 5 in the rounding (tenths) place. 0.574

Since 7 is greater than 5, change the 5 to a 6. 0.574

Drop the digits to the right of the tenths place. 0.6

1. Identify the number in the rounding place.
2. Look at the digit to its right.
3. If the digit is 5 or greater, increase the number in the rounding place by 1. If the digit is 4 or less, keep the number in the rounding place the same.
4. Drop all digits to the right of the rounding place.

Example: Round 2.783 to the nearest hundredth.

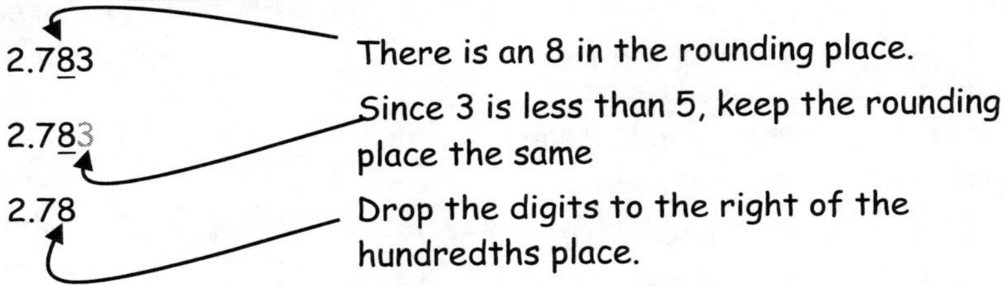

2.783
There is an 8 in the rounding place.
2.783
Since 3 is less than 5, keep the rounding place the same
2.78
Drop the digits to the right of the hundredths place.

Adding and subtracting decimals is very similar to adding or subtracting whole numbers. The main difference is that you have to line-up the decimal points in the numbers before you begin.

Examples: Find the sum of 3.14 and 1.2.

$$\begin{array}{r} 3.14 \\ +\,1.20 \\ \hline 4.34 \end{array}$$

1. Line up the decimal points. Add zeroes as needed.
2. Add (or subtract) the decimals.
3. Add (or subtract) the whole numbers.
4. Bring the decimal point straight down.

Add 55.1, 6.472 and 18.33.

$$\begin{array}{r} 55.100 \\ 6.472 \\ +\,18.330 \\ \hline 79.902 \end{array}$$

Examples: Subtract 3.7 from 9.3.

$$\begin{array}{r} 9.3 \\ -\,3.7 \\ \hline 5.6 \end{array}$$

Find the difference of 4.1 and 2.88.

$$\begin{array}{r} 4.10 \\ -\,2.88 \\ \hline 1.22 \end{array}$$

Help Pages

Solved Examples

Decimals (continued)

When **multiplying a decimal by a whole number**, the process is similar to multiplying whole numbers.

Examples: Multiply 3.42 by 4.

```
  3.42 —— 2 decimal places
 ×   4 —— 0 decimal places
 13.68 —— Place decimal point
          so there are 2
          decimal places.
```

1. Line up the numbers on the right.
2. Multiply. Ignore the decimal point.
3. Place the decimal point in the product. (The total number of decimal places in the product must equal the total number of decimal places in the factors.)

Find the product of 2.3 and 2.

```
  2.3 —— 1 decimal place
 ×  2 —— 0 decimal places
  4.6 —— Place decimal point
         so there is 1
         decimal place.
```

The process for **multiplying two decimal numbers** is a lot like the process described above.

Examples: Multiply 0.4 by 0.6.

```
   0.4 —— 1 decimal place
 × 0.6 —— 1 decimal place
  0.24 —— Place decimal point
          so there are 2
          decimal places.
```

Find the product of 2.67 and 0.3.

```
  2.67 —— 2 decimal places
 × 0.3 —— 1 decimal place
 0.801 —— Place decimal point
          so there are 3
          decimal places.
```

Sometimes it is necessary to add **zeroes in the product** as placeholders in order to have the correct number of decimal places.

Example: Multiply 0.03 by 0.4.

```
  0.03 —— 2 decimal places
 × 0.4 —— 1 decimal place
 0.012 —— Place decimal point
          so there are 3
          decimal places.
```

A zero had to be added in front of the 12 so there would be 3 decimal places in the product.

The process for **dividing a decimal number by a whole number** is similar to dividing whole numbers.

Examples: Divide 6.4 by 8.

```
    0.8
 8)6.4
  -6 4
     0
```

1. Set up the problem for long division.
2. Place the decimal point in the quotient directly above the decimal point in the dividend.
3. Divide. Add zeroes as placeholders if necessary. (examples below)

Find the quotient of 20.7 and 3.

```
    6.9
 3)20.7
  -18
    27
   -27
     0
```

Examples: Divide 4.5 by 6.

```
   0.75
 6)4.50  ← Add zero(es).
  -42↓
    30   ← Bring zero down.
   -30
     0
```

Find the quotient of 3.5 and 4.

```
   0.875
 4)3.500  ← Add zero(es).
  -3 2↓↓
     30↓  ← Bring zero down.
    -28↓
      20  ← Bring zero down.
     -20
       0
```

Add zero(es).

Bring zero down.

Keep dividing.

Help Pages

Solved Examples

Decimals (continued)

When dividing decimals, the remainder is not always zero. Sometimes, the division continues on and on and the remainder begins to repeat itself. When this happens the quotient is called a **repeating decimal**.

Examples: Divide 2 by 3. Divide 10 by 11.

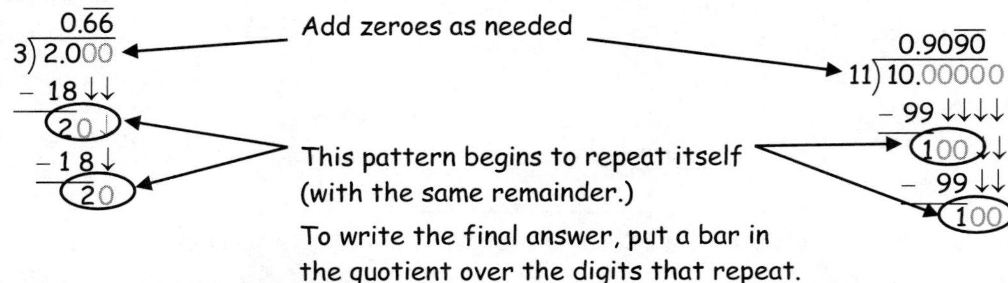

To write the final answer, put a bar in the quotient over the digits that repeat.

The process for **dividing a decimal number by a decimal number** is similar to other long division that you have done. The main difference is that the decimal point has to be moved in both the dividend and the divisor the same number of places to the right.

Examples: Divide 1.8 by 0.3. Divide 0.385 by 0.05.

```
      6.
0.3)1.8
   −18
     0
```

1. Change the divisor to a whole number by moving the decimal point as many places to the right as possible.
2. Move the decimal in the dividend the same number of places to the right as you did in the divisor.
3. Put the decimal point in the quotient directly above the decimal point in the dividend.
4. Divide.

```
         7.7
0.05)0.385
     −35↓
       35
      −35
        0
```

Geometry

Finding the **area of a parallelogram** is similar to finding the area of any other quadrilateral. The area of the figure is equal to the length of its base multiplied by the height of the figure.

Area of parallelogram = base × height or $A = b \times h$

Example: Find the area of the parallelogram below.

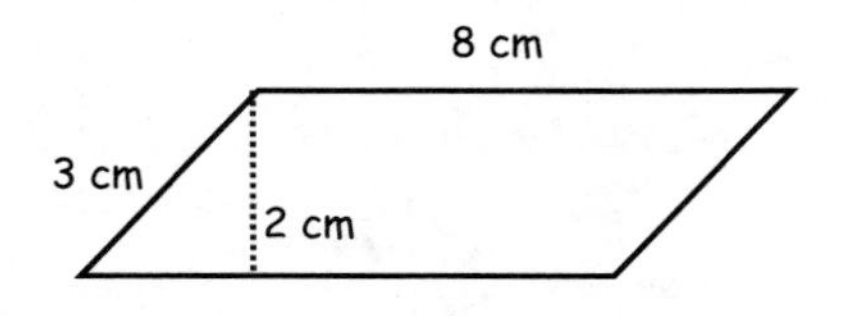

1. Find the length of the base. (8 cm)
2. Find the height. (It is 2cm. The height is always straight up and down – never slanted.)
3. Multiply to find the area. (16 cm^2)

So, $A = 8\text{ cm} \times 2\text{ cm} =$ **16 cm^2**.

Help Pages

Solved Examples

Geometry (continued)

To find the **area of a triangle**, it is helpful to recognize that any triangle is exactly half of a parallelogram.

The whole figure is a parallelogram.

Half of the whole figure is a triangle.

So, the triangle's area is equal to half of the product of the base and the height.

$$\text{Area of triangle} = \frac{1}{2}(\text{base} \times \text{height}) \quad \text{or} \quad A = \frac{1}{2}bh$$

Examples: Find the area of the triangles below.

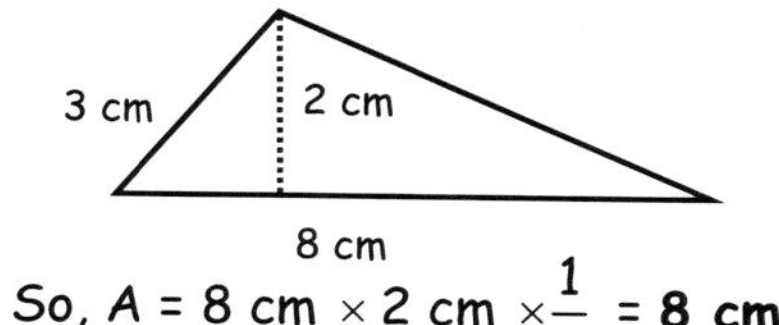

So, $A = 8\text{ cm} \times 2\text{ cm} \times \frac{1}{2} = \mathbf{8\ cm^2}$.

1. Find the length of the base. (8 cm)
2. Find the height. (It is 2cm. The height is always straight up and down – never slanted.)
3. Multiply them together and divide by 2 to find the area. (8 cm^2)

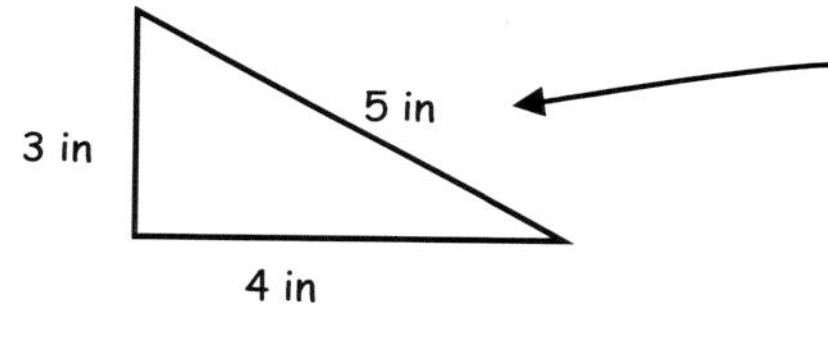

The base of this triangle is 4 inches long. Its height is 3 inches. (Remember the height is always straight up and down!)

So, $A = 4\text{ in} \times 3\text{ in} \times \frac{1}{2} = \mathbf{6\ in^2}$

Finding the **area of a trapezoid** is a little different than the other quadrilaterals that we have seen. Trapezoids have 2 bases of unequal length. To find the area, first find the average of the lengths of the 2 bases. Then, multiply that average by the height.

$$\text{Area of trapezoid} = \frac{\text{base}_1 + \text{base}_2}{2} \times \text{height} \quad \text{or} \quad A = \left(\frac{b_1 + b_2}{2}\right)h$$

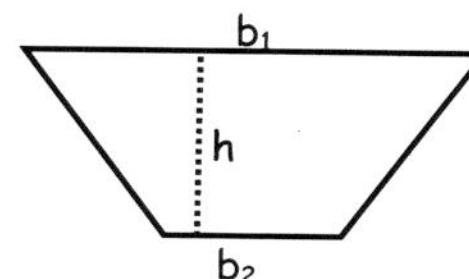

The bases are labeled b_1 and b_2.

The height, h, is the distance between the bases.

Example: Find the area of the trapezoid below.

1. Add the lengths of the two bases. (22 cm)
2. Divide the sum by 2. (11 cm)
3. Multiply that result by the height to find the area. (110 cm^2)

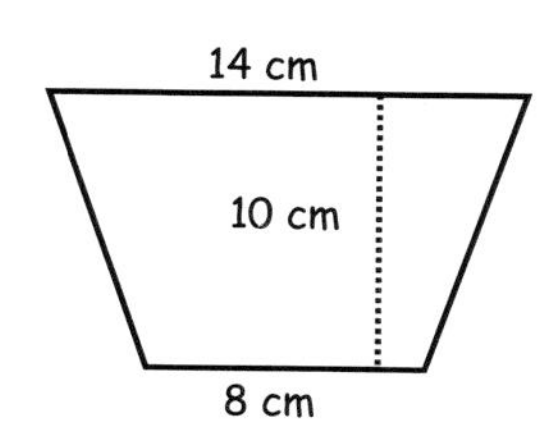

$$\frac{14\text{cm} + 8\text{cm}}{2} = \frac{22\text{cm}}{2} = 11\text{cm}$$

$11\text{ cm} \times 10\text{ cm} = \mathbf{110\ cm^2} = \text{Area}$

Help Pages

Solved Examples

Geometry (continued)

The **circumference of a circle** is the distance around the outside of the circle. Before you can find the circumference of a circle you must know either its radius or its diameter. Also, you must know the value of the constant, pi (π). $\pi = 3.14$ (rounded to the nearest hundredth).

Once you have this information, the circumference can be found by multiplying the diameter by pi.

Circumference = $\pi \times$ diameter or $C = \pi d$

Examples: Find the circumference of the circles below.

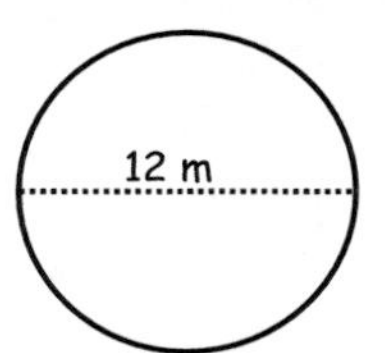

1. Find the length of the diameter. (12 m)
2. Multiply the diameter by π. (12m $\times$ 3.14).
3. The product is the circumference. (37.68 m)

So, C = 12 m $\times$ 3.14 = **37.68 m.**

Sometimes the radius of a circle is given instead of the diameter. Remember, the radius of any circle is exactly half of the diameter. If a circle has a radius of 3 feet, its diameter is 6 feet.

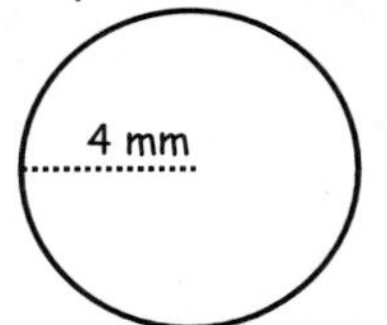

Since the radius is 4 mm, the diameter must be 8 mm.

Multiply the diameter by π. (8 mm $\times$ 3.14).

The product is the circumference. (25.12 mm)

So, C = 8 mm $\times$ 3.14 = **25.12 mm.**

When finding the **area of a circle**, the length of the radius is squared (multiplied by itself), and then that answer is multiplied by the constant, pi (π). $\pi = 3.14$ (rounded to the nearest hundredth).

Area = $\pi \times$ radius $\times$ radius or $A = \pi r^2$

Examples: Find the area of the circles below.

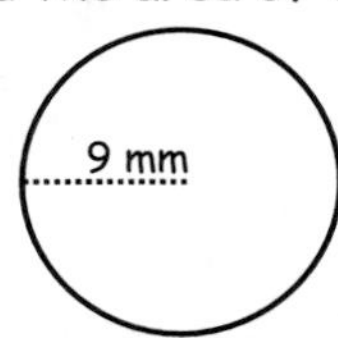

1. Find the length of the radius. (9 mm)
2. Multiply the radius by itself. (9 mm x 9 mm)
3. Multiply the product by pi. (81 mm^2 x 3.14)
4. The result is the area. (254.34 mm^2)

So, A = 9 mm x 9 mm x 3.14 = **254.34 mm^2.**

Sometimes the diameter of a circle is given instead of the radius. Remember, the diameter of any circle is exactly twice the radius. If a circle has a diameter of 6 feet, its radius is 3 feet.

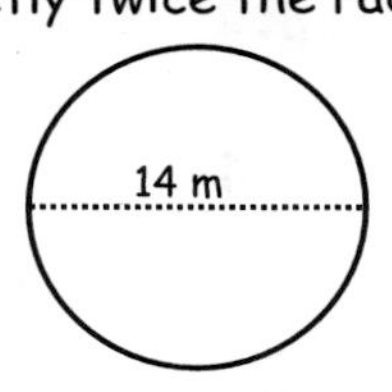

Since the diameter is 14 m, the radius must be 7 m.

Square the radius. (7 m x 7 m)

Multiply that result by π. (49 m^2 $\times$ 3.14).

The product is the area. (153.86 m^2)

So, $A = (7\text{ m})^2 \times 3.14$ = **153.86 m^2.**

Help Pages

Solved Examples

Geometry (continued)

To find the **surface area** of a solid figure, it is necessary to first count the total number of faces. Then, find the area of each of the faces; finally, add the areas of each face. That sum is the surface area of the figure.

Here, the focus will be on finding the surface area of a rectangular prism. A rectangular prism has 6 faces. Actually, the opposite faces are identical, so this figure has 3 pairs of faces. Also, a prism has only 3 dimensions: Length, Width, and Height.

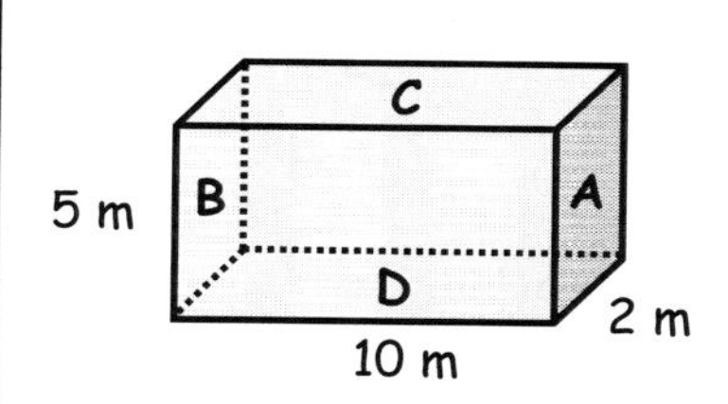

This prism has identical left & right sides (A & B), identical top and bottom (C & D), and identical front and back (unlabeled).

1. Find the area of the front: L x W. (10 m x 5 m = 50 m^2) Since the back is identical, its area is the same.
2. Find the area of the top (C): L x H. (10 m x 2 m = 20 m^2) Since the bottom (D) is identical, its area is the same.
3. Find the area of side A: W x H. (2 m x 5 m = 10 m^2) Since side B is identical, its area is the same.
4. Add up the areas of all 6 faces. (10 m^2 + 10 m^2 + 20 m^2 + 20 m^2 + 50 m^2 + 50 m^2 = **160 m^2**)

The formula is Surface Area = 2(length x width) + 2(length x height) + 2(width x height)

or $SA = 2LW + 2LH + 2WH$

Ordering Integers

Integers include the counting numbers, their opposites (negative numbers) and zero.

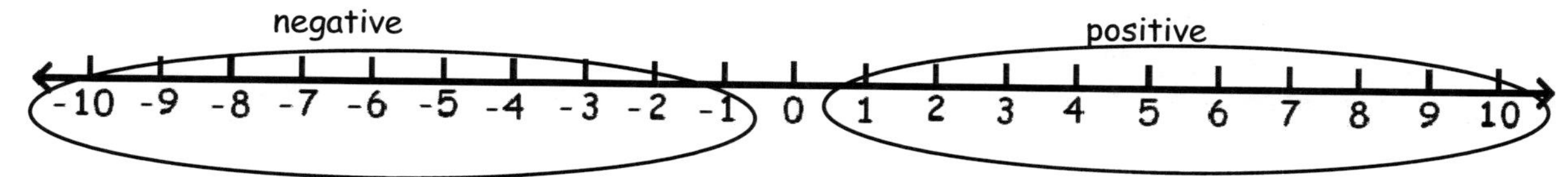

The negative numbers are to the left of zero. The positive numbers are to the right of zero.

The further a number is to the right, the greater its value. For example, 9 is further to the right than 2, so 9 is greater than 2.

In the same way, -1 is further to the right than -7, so -1 is greater than -7.

Examples: Order these integers from **least to greatest**: -10, 9, -25, 36, 0

Remember, the smallest number will be the one farthest to the left on the number line, -25, then -10, then 0. Next will be 9, and finally 36.

Answer: -25, -10, 0, 9, 36

Order these integers from **least to greatest**: -44, -19, -56, -80, -2

Answer: -80, -56, -44, -19, -2 (-80 is farthest to the left, so it is smallest. -2 is farthest to the right, so it is greatest.)

Put these integers in order from **greatest to least**: -94, -6, -24, -70, -14

Now the greatest value (farthest to the right) will come first and the smallest value (farthest to the left) will come last.

Answer: -6, -14, -24, -70, -94

Help Pages

Solved Examples

Ratio and Proportion

A **ratio** is used to compare two numbers. There are three ways to write a ratio comparing 5 and 7:

1. Word form ➡ 5 to 7
2. Fraction form ➡ $\frac{5}{7}$
3. Ratio form ➡ 5 : 7

All are pronounced "five to seven."

You must make sure that all ratios are written in simplest form. (Just like fractions!!)

A **proportion** is a statement that two ratios are equal to each other. There are two ways to solve a proportion when a number is missing.

1. One way to solve a proportion is already familiar to you. You can use the equivalent fraction method.

$$\frac{5}{8} = \frac{n}{64} \quad (\times 8)$$

$n = 40$

So, $\frac{5}{8} = \frac{40}{64}$.

To use Cross-Products:

1. Multiply downward on each diagonal.
2. Make the product of each diagonal equal to each other.
3. Solve for the missing variable.

2. Another way to solve a proportion is by using cross-products.

$$\frac{14}{20} = \frac{21}{n}$$

$$20 \times 21 = 14 \times n$$

$$420 = 14n$$

$$\frac{420}{14} = \frac{14n}{14}$$

$$30 = n$$

So, $\frac{14}{20} = \frac{21}{30}$.

Percent

When changing from a fraction to a percent, a decimal to a percent, or from a percent to either a fraction or a decimal, it is very helpful to use an FDP chart (Fraction, Decimal, Percent).

To change a **fraction to a percent and/or decimal**, first find an equivalent fraction with 100 in the denominator. Once you have found that equivalent fraction, it can easily be written as a decimal. To change that decimal to a percent, move the decimal point 2 places to the right and add a % sign.

Example: Change $\frac{2}{5}$ to a percent and then to a decimal.

1. Find an equivalent fraction with 100 in the denominator.
2. From the equivalent fraction above, the decimal can easily be found. Say the name of the fraction: "forty hundredths." Write this as a decimal: 0.40.
3. To change 0.40 to a percent, move the decimal two places to the right. Add a % sign.

F	D	P
$\frac{2}{5}$		

F	D	P
$\frac{2}{5} = \frac{?}{100}$	0.40	

F	D	P
$\frac{2}{5} = \frac{40}{100}$	0.40	40%

$$\frac{2}{5} = \frac{?}{100} \quad (\times 20)$$

? = 40

$$\frac{2}{5} = \frac{40}{100} = 0.40$$

0.40 = 40%

Help Pages

Solved Examples

Percent (continued)

When changing from a **percent to a decimal or a fraction**, the process is similar to the one used on the previous page. Begin with the percent. Write it as a fraction with a denominator of 100; reduce this fraction. Return to the percent, move the decimal point 2 places to the left. This is the decimal.

Example: Write 45% as a fraction and then as a decimal.

1. Begin with the percent. (45%) Write a fraction where the denominator is 100 and the numerator is the "percent." $\frac{45}{100}$
2. This fraction must be reduced. The reduced fraction is $\frac{9}{20}$.
3. Go back to the percent. Move the decimal point two places to the left to change it to a decimal.

F	D	P
$\frac{45(\div 5)}{100(\div 5)} = \frac{9}{20}$		45%

F	D	P
$\frac{9}{20}$		45% = 0.45

F	D	P
$\frac{9}{20}$	0.45	45%

When changing from a **decimal to a percent or a fraction**, again, the process is similar to the one used above. Begin with the decimal. Move the decimal point 2 places to the right and add a % sign. Return to the decimal. Write it as a fraction and reduce.

Example: Write 0.12 as a percent and then as a fraction.

1. Begin with the decimal. (0.12) Move the decimal point two places to the right to change it to a percent.
2. Go back to the decimal and write it as a fraction. Reduce this fraction.

F	D	P
	0.12	

F	D	P
	0.12 = 12%	12%

F	D	P
$\frac{12(\div 4)}{100(\div 4)} = \frac{3}{25}$	0.12	12%

Help Pages

Solved Examples

Compound Probability

The **probability of two or more independent events** occurring together can be determined by multiplying the individual probabilities together. The product is called the compound probability.

Probability of A & B = (Probability of A) x (Probability of B)

or P(A and B) = P(A) x P(B)

Example: What is the probability of rolling a 6 and then a 2 on two rolls of a die [P(6 and 2)]?

A) First, find the probability of rolling a 6 [P(6)]. Since there are 6 numbers on a die and only one of them is a 6, the probability of getting a 6 is $\frac{1}{6}$.

B) Then find the probability of rolling a 2 [P(2)]. Since there are 6 numbers on a die and only one of them is a 2, the probability of getting a 2 is $\frac{1}{6}$.

So, P(6 and 2) = P(6) x P(2) = $\frac{1}{6} \times \frac{1}{6} = \frac{1}{36}$.

There is a 1 to 36 chance of getting a 6 and then a 2 on two rolls of a die.

Example: What is the probability of getting a 4 and then a number greater than 2 on two spins of this spinner [P(4 and greater than 2)]?

A) First, find the probability of getting a 4 [P(4)]. Since there are 4 numbers on the spinner and only one of them is a 4, the probability of getting a 4 is $\frac{1}{4}$.

B) Then find the probability of getting a number greater than 2 [P(greater than 2)]. Since there are 4 numbers on the spinner and two of them are greater than 2, the probability of getting a 2 is $\frac{2}{4}$.

So, P(2 and greater than 2) = P(2) x P(greater than 2) = $\frac{1}{4} \times \frac{2}{4} = \frac{2}{16} = \frac{1}{8}$.

There is a 1 to 8 chance of getting a 4 and then a number greater than 2 on two spins of a spinner.

Example: On three flips of a coin, what is the probability of getting heads, tails, heads [P(H,T,H)]?

A) First, find the probability of getting heads [P(H)]. Since there are only 2 sides on a coin and only one of them is heads, the probability of getting heads is $\frac{1}{2}$.

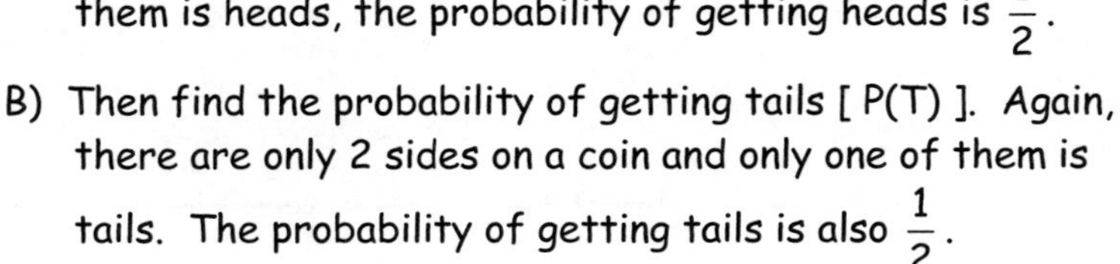

B) Then find the probability of getting tails [P(T)]. Again, there are only 2 sides on a coin and only one of them is tails. The probability of getting tails is also $\frac{1}{2}$.

So, P(H,T,H) = P(H) x P(T) x P(H) = $\frac{1}{2} \times \frac{1}{2} \times \frac{1}{2} = \frac{1}{8}$.

There is a 1 to 8 chance of getting heads, tails and then heads on 3 flips of a coin.

Who Knows???

Degrees in a right angle? .... (90°)
A straight angle? ....(180°)
Angle greater than 90°? .... (obtuse)
Less than 90°? ....(acute)
Sides in a quadrilateral? .... (4)
Sides in an octagon? .... (8)
Sides in a hexagon? .... (6)
Sides in a pentagon? .... (5)
Sides in a heptagon? .... (7)
Sides in a nonagon? .... (9)
Sides in a decagon?.......................(10)
Inches in a yard? ....(36)
Yards in a mile? ....(1,760)
Feet in a mile? ....(5,280)
Centimeters in a meter? ....(100)
Teaspoons in a tablespoon? .... (3)
Ounces in a pound? ....(16)
Pounds in a ton? ....(2,000)
Cups in a pint? .... (2)
Pints in a quart? .... (2)
Quarts in a gallon? .... (4)
Millimeters in a meter? ....(1,000)
Years in a century? ....(100)
Years in a decade? ....(10)
Celsius freezing? .... (0°C)
Celsius boiling? ....(100°C)
Fahrenheit freezing? ....(32°F)

Fahrenheit boiling? .... (212°F)
Number with only 2 factors? ....(prime)
Perimeter? ....(add the sides)
Area?..................................(length x width)
Volume? .... (length x width x height)
Area of parallelogram? ......(base x height)
Area of triangle? .... ($\frac{1}{2}$ base x height)
Area of trapezoid ... ($\frac{\text{base} + \text{base}}{2}$ x height)
Surface Area of a Rectangular Prism? .... 2(LW) + 2(WH) + 2(LH)
Area of a circle? .... (πr^2)
Circumference of a circle? .... ($d\pi$)
Triangle with no sides equal? ...... (scalene)
Triangle with 3 sides equal? .. (equilateral)
Triangle with 2 sides equal? ..... (isosceles)
Distance across the middle of a circle? .... (diameter)
Half of the diameter? ....(radius)
Figures with the same size and shape? ....(congruent)
Figures with same shape, different sizes? ....(similar)
Number occurring most often? ...... (mode)
Middle number? .... (median)
Answer in addition? ....(sum)
Answer in division? ....(quotient)
Answer in multiplication? ....(product)
Answer in subtraction?..........(difference)

Intermediate B

Mathematics

Answers to Lessons

Lesson #1		Lesson #2		Lesson #3	
1	$2\frac{2}{3}$	**1**	0.454	**1**	0.60; 60%
2	0.0204	**2**	9	**2**	14 r 12
3	75	**3**	2.12	**3**	2
4	seven thousand five hundred thirty-four ten thousandths	**4**	$\frac{1}{4}$; $\frac{3}{4}$	**4**	12,978
5	24,219	**5**	$\frac{2}{3}$ or 2 to 3	**5**	25.12 in.
6	$14\frac{13}{15}$	**6**	15,266	**6**	$13\frac{3}{20}$
7	112 cm^2	**7**	9:45	**7**	8 ft.
8	1,309	**8**	$x = 48$	**8**	1,000 g
9	10	**9**	38,000,000	**9**	22,304
10	0.35; 35%	**10**	70,776	**10**	125 yd.
11	1, 2, 3, 4, 6, 8, 12, 24	**11**	isosceles	**11**	4.05
12	$x = 39$	**12**	42	**12**	$x = 75$
13	186,018	**13**	65 inches	**13**	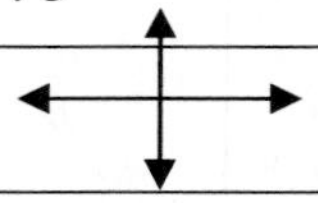
14	47,900,000	**14**	1,333	**14**	2,006
15	0.035; 0.3; 0.503; 0.53	**15**	$7\frac{3}{4}$	**15**	$4\frac{2}{5}$
16	2/3	**16**	24 cm^2	**16**	21,120
17	40	**17**	<	**17**	five thousand seven hundred thirty-two ten thousandths
18	30%	**18**	3,145	**18**	GCF = 4; LCM = 80
19	61	**19**	3/5	**19**	7,000 lb.
20	scalene	**20**	triangle; (1, 4)	**20**	1/2 in.

Lesson #4		Lesson #5		Lesson #6	
1	70 years	**1**	$a = 6$	**1**	25
2	90.38	**2**	$3\frac{1}{2}$	**2**	addition
3	34,223	**3**	8,025	**3**	$7\frac{2}{9}$
4	4.762; 4.7; 4.267; 4.076	**4**	2/3	**4**	
5	21	**5**	50,184	**5**	90
6	12 r 21	**6**	34.35	**6**	congruent
7	1	**7**	33.97	**7**	2/3
8	78 in^2	**8**	$1\frac{1}{2}$	**8**	18,204
9	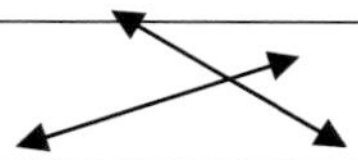	**9**	216 m^2	**9**	2/3
10	15.36	**10**	0.00105	**10**	40,000,000
11	2	**11**	0.28; 28%	**11**	$\frac{21}{50}$; 42%
12	6:45	**12**	9:15	**12**	0
13	$14\frac{9}{20}$	**13**	$<$	**13**	$n = 24$
14	35%; $\frac{7}{20}$	**14**	seven and one thousand three hundred ninety-two ten thousandths	**14**	$3\frac{6}{7}$
15	equilateral	**15**	8,800 yards	**15**	30 ft^2
16	475 r 8	**16**	$10\frac{11}{15}$	**16**	26.05
17	1,980 yd.	**17**		**17**	24 tsp.
18	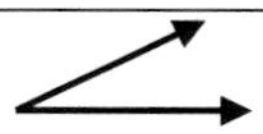	**18**	182,261	**18**	
19	3	**19**	$2^3 \times 3^2$	**19**	1,033,186
20	38°F	**20**	65 cups	**20**	10

Lesson #7		Lesson #8		Lesson #9	
1	108	1	3,000 mm	1	$1.49
2	120	2	$2\frac{2}{3}$	2	33,444
3	100°C	3	$C = d\pi$ or $2\pi r$	3	32 cm
4	$\frac{18}{5}$	4	1,758	4	20
5	595,595	5	fifty-seven and two hundred thirty-four thousandths	5	900 cm
6	12.56 ft^2	6	9	6	87,000
7	$5\frac{5}{9}$	7	7	7	17.07
8	6.04	8	$\frac{4}{3}$	8	1,308,921
9	polygons	9	30 cupcakes	9	both are 47
10	5,556	10	46.1	10	0.0835
11	78	11	prime	11	180°
12	>	12	69,225	12	4.003; 4.035; 4.3; 4.305
13	36 in.	13	$x = 24$	13	1:52
14		14	1.4	14	6
15	$6.00	15	$3\frac{3}{7}$	15	
16	0.00054	16	86.27	16	2/3
17	mode	17	144 in^2	17	0.15; 15%
18	295	18	acute	18	$13\frac{11}{24}$
19	quotient	19	$\frac{6}{5}$	19	1/4
20	2/3	20	8,000 lb.	20	108,544

Lesson #10		Lesson #11		Lesson #12	
1	1,309	**1**	5	**1**	0.12; $\frac{3}{25}$
2	$2\frac{3}{8}$ in.	**2**	0.36; $\frac{9}{25}$	**2**	$x = 24$
3	0.141	**3**	10.5	**3**	15
4	3	**4**	3.98	**4**	9 ft.
5	46.46	**5**	$3\frac{7}{9}$	**5**	0.3652
6	1,101	**6**	5 ft.	**6**	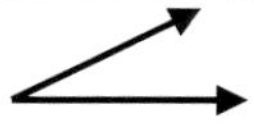
7	circumference	**7**	284.28	**7**	$15\frac{13}{21}$
8	$4\frac{3}{4}$	**8**	27,432	**8**	isosceles
9	35%	**9**	$1\frac{4}{5}$	**9**	1,000
10	4	**10**	1,000 g	**10**	3
11	4,410	**11**	1/6	**11**	15.70 inches
12	$x = 15$	**12**	41,266	**12**	1/3
13	$\frac{20}{3}$	**13**	$15\frac{13}{20}$	**13**	3.027
14	36 months	**14**	32 trucks	**14**	32
15	prime number	**15**	27	**15**	190 mm
16	12	**16**	28	**16**	$<$
17	liters	**17**	2/5	**17**	180°
18	1,676	**18**	4.8	**18**	39 sparrows
19	7 tons	**19**	0.0018	**19**	14,552
20	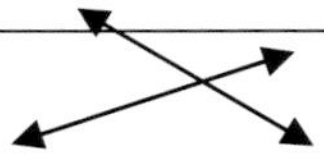	**20**	12:30	**20**	prime

Lesson #13		Lesson #14		Lesson #15	
1	20,000	**1**	144 ft^3	**1**	90°
2	$2\frac{2}{5}$	**2**	18	**2**	1,479,081
3	66 in^2	**3**	144 in.	**3**	heptagon
4	34,309	**4**	20.60	**4**	1/4
5	212°F	**5**	0.45; 45%	**5**	<
6	0.0237	**6**	3.2	**6**	1/2
7	12.02	**7**	67 in.	**7**	128 cm^2
8	1,818,718	**8**	0.00028	**8**	27,000
9	15 tsp.	**9**	65 lions	**9**	51
10	<	**10**	$15\frac{11}{15}$	**10**	7
11	69,012	**11**	34 r 15	**11**	$\frac{30}{7}$
12		**12**	5/8; 5 to 8	**12**	3,677
13	3	**13**	1/2	**13**	parallelogram
14	57.8	**14**	1,018,904	**14**	3.51
15	$SA = 2LW + 2LH + 2WH$	**15**	79	**15**	product
16	5 lb.	**16**	20	**16**	$A = \pi r^2$
17	prime number	**17**	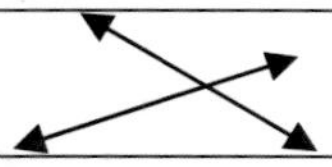	**17**	14.7
18	eight and one thousand seven hundred forty-two ten thousandths	**18**	$A = \frac{1}{2}bh$	**18**	Yes
19	$3\frac{1}{3}$	**19**	$1\frac{1}{8}$ in.	**19**	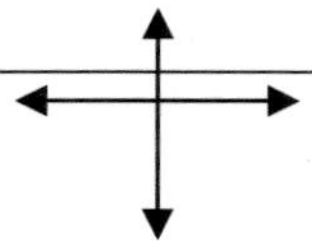
20	30 students	**20**	2,387	**20**	20 pieces

Lesson #16		Lesson #17		Lesson #18	
1	GCF = 4 LCM = 24	**1**	12 pizzas	**1**	15%; $\frac{3}{20}$
2	$x = 70$	**2**	2/3	**2**	$x = 9$
3	2, 3, 5, 7	**3**	17,153	**3**	nonagon
4	0.40; 40%	**4**	$10\frac{1}{2}$	**4**	$1\frac{4}{5}$
5	1,000 mm	**5**	124.3	**5**	16,524
6	congruent	**6**	0.25; $\frac{1}{4}$	**6**	87
7	1/2	**7**	21 r 1	**7**	prime
8	difference	**8**	polygons	**8**	4,000 mm
9	$2\frac{1}{5}$	**9**	$>$	**9**	5:00
10	10,560 ft.	**10**	two and eight thousand six hundred seventy-five ten thousandths	**10**	53,358
11	$12\frac{23}{30}$	**11**	8/9	**11**	81 ft^2
12	2.075; 2.5; 2.705; 2.75	**12**	30 ft^2	**12**	0.192
13	24.595	**13**	$12\frac{17}{21}$	**13**	15%
14	90	**14**	32°F	**14**	circumference
15	$<$	**15**	4,371	**15**	16
16	$3^2 \times 5$	**16**	quotient	**16**	1.76; 1.67; 1.607; 1.076
17	7:30	**17**	$x = 21$	**17**	scalene
18	$3\frac{1}{2}$	**18**	1,760 yd.	**18**	7,040 yards
19	0.000064	**19**	16,257,583	**19**	0.060
20	12,000 lb.	**20**		**20**	$13\frac{13}{40}$

	Lesson #19		Lesson #20		Lesson #21
1	2, 3, 5, 7, 11	**1**	62.02	**1**	3
2	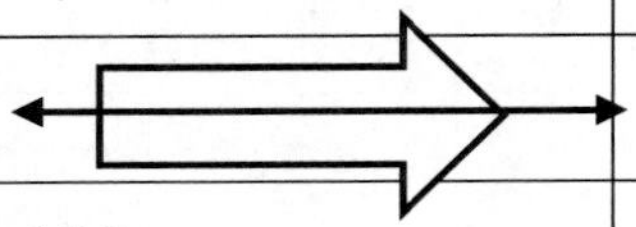	**2**	15	**2**	8:50
3	108	**3**	1/2	**3**	1/2
4	75¢	**4**	cylinder	**4**	90
5	2^6	**5**	8,275	**5**	46.29
6	16.05	**6**	49 in^2	**6**	983
7	170,875	**7**	1	**7**	0.65; $\frac{13}{20}$
8	40	**8**	4,000	**8**	700 years
9	$3\frac{4}{7}$	**9**	0.0535	**9**	4.00
10	88,626	**10**	$\frac{6}{7}$; 6:7	**10**	30 in^2
11	96 oz.	**11**	295	**11**	$7\frac{3}{7}$
12	87,000,000	**12**	24 cm^2	**12**	prime number
13	$<$	**13**	$14\frac{11}{15}$	**13**	$>$
14	8 C	**14**	720	**14**	product
15	decagon	**15**	15,840 ft.	**15**	4/5
16	9 ft.	**16**	$x = 45$	**16**	2.9
17	8	**17**	0.35; 35%	**17**	2 pt.
18	15.60	**18**	12 weeks	**18**	4/5
19	$\frac{19}{3}$	**19**	2	**19**	similar
20	8:30 a.m.; 6°	**20**		**20**	equilateral

Lesson #22		Lesson #23		Lesson #24	
1	4,000 mm	**1**	7,419	**1**	5
2	$>$	**2**	$13\frac{5}{9}$	**2**	112 daisies
3	36	**3**	45.8; 45.809; 45.908; 45.98	**3**	0.00210
4	0.000032	**4**	$\frac{44}{5}$	**4**	$8\frac{1}{4}$
5	$\frac{7}{5}$	**5**	$<$	**5**	fifty-six and seven tenths
6	1, 2, 3, 6, 9, 18	**6**	5,900,000	**6**	22.12
7	$1\frac{1}{3}$	**7**	1,034	**7**	
8	48 cans	**8**	isosceles	**8**	$>$
9	35	**9**	kilometer	**9**	100°C
10	milliliters	**10**	444,737	**10**	573
11	60 qt.	**11**	55%; $\frac{11}{20}$	**11**	3,520 yd.
12	144	**12**	$x = 45$	**12**	2,435
13		**13**	0.435	**13**	68 in.
14	2/3	**14**	6,000 lb.	**14**	11:52 a.m.
15	six and twenty-one hundredths	**15**	2/3	**15**	1,602 r 5
16	radius	**16**	1, 3, 5, 15	**16**	$\frac{17}{3}$
17	2,231	**17**	31,801	**17**	$10\frac{17}{30}$
18	7	**18**	120	**18**	1/5
19	18 r 1	**19**	24 tsp.	**19**	
20	0.06; 6%	**20**	42 in.	**20**	1/2; 1/6

Lesson #25		Lesson #26		Lesson #27	
1	0.693	**1**	416	**1**	3.007
2	$1\frac{5}{7}$	**2**	222 mm^2	**2**	$3^2 \times 5$
3	1/4	**3**	>	**3**	40 cm
4	<	**4**	21	**4**	$2\frac{2}{3}$
5	91.38	**5**	13,981	**5**	14
6	144	**6**	5	**6**	271,966
7	mode	**7**	10,665	**7**	500 cm
8	328	**8**	composite number	**8**	0.00028
9	21.98 in.	**9**		**9**	3 lb.
10	20	**10**	6 C	**10**	13,569
11	35	**11**	17,632	**11**	14,000 lb.
12	6 m	**12**	25.003	**12**	0.45; 45%
13	0.75; $\frac{3}{4}$	**13**	56	**13**	0.485
14	$2^3 \times 3$	**14**	15	**14**	$A = \frac{1}{2}bh$
15	polygons	**15**	sum	**15**	10
16	$5\frac{3}{4}$	**16**	0.24; 24%	**16**	7
17	0°C; 32°F	**17**	115 ft^2	**17**	$\frac{39}{5}$
18	$1\frac{1}{4}$	**18**	$8\frac{2}{5}$	**18**	quotient
19	8 decades	**19**	Yes	**19**	25 r 7
20	7,000 g	**20**	2^5	**20**	congruent

Lesson #28		Lesson #29		Lesson #30	
1	diameter	**1**	90 knives	**1**	equilateral
2	8:30	**2**	21,677	**2**	100°C; 212°F
3	4,806	**3**	25%; $\frac{1}{4}$	**3**	59
4	12	**4**	0.3105	**4**	3,154
5	7 cm	**5**	78 m^3	**5**	96 in.
6	4,855	**6**	$\frac{5}{6}$	**6**	$\frac{2}{3}$
7	2/5	**7**	1, 3, 7, 21	**7**	12 cm^3
8	23.9	**8**	110,000	**8**	110 r 3
9	$\frac{4}{3}$	**9**		**9**	24 qt.
10	$<$	**10**	3	**10**	$2^3 \times 7$
11	9,343,114	**11**	6:20	**11**	$14\frac{5}{9}$
12	$\frac{27}{5}$	**12**	242,670	**12**	A = bh
13	75	**13**	$<$	**13**	7:45
14	121.6	**14**	GCF = 4 LCM = 48	**14**	obtuse
15	2	**15**	nonagon	**15**	132.1
16	$8\frac{1}{4}$	**16**	1,760 yd.	**16**	$1\frac{1}{3}$
17	right angle	**17**	12	**17**	16 pt.
18	16.1	**18**	$x = 48$	**18**	$10\frac{5}{9}$
19	26,400 ft.	**19**	5.86; 5.8; 5.668; 5.608	**19**	49
20	$\frac{1}{2}, \frac{3}{4}$	**20**	128,390	**20**	85%; $\frac{17}{20}$